Urban Transportation and the Environment

Sudhakar Yedla

Urban Transportation and the Environment

Issues, Alternatives and Policy Analysis

Springer

Sudhakar Yedla
Indira Gandhi Institute of Development Research
Goregaon, Mumbai, India

ISBN 978-81-322-2312-2 ISBN 978-81-322-2313-9 (eBook)
DOI 10.1007/978-81-322-2313-9

Library of Congress Control Number: 2015936342

Springer New Delhi Heidelberg New York Dordrecht London

Printed on acid-free paper

Springer (India) Pvt. Ltd. is part of Springer Science+Business Media (www.springer.com)

"To my loving son Dhiren"

Preface

Small towns grow up riding bicycle both for short daily trips and long and occasional trips on "ride-cycle-ride" mode. The small towns and villages in India have always been sustainable, whether it is transportation, or waste management, or even resources management. It is only a surprise that, in the name of green growth, the international development agencies are now trying to introduce the same old principles of sustainability to the developing world which they have already been following since long (it is even practiced today in smaller towns/villages).

Rapid economic growth brings opportunities and they bring significant changes in lifestyles. The average distance between "home" and the office has increased (as we call it migration) and so have the daily trip lengths. Due to the fastness in movements and longer distances to travel in shorter time, motorization has become the face of transportation in Indian metropolitan cities. Services sector-driven economic growth has resulted in increased per capita incomes and purchasing power. With the need to move faster and farther, the number of personal vehicles (cars and bikes) has increased many times over the last two to three decades. The infrastructure provision also got skewed towards motorized modes, marginalizing the golden and age-old mode of transport—the non-motorized modes such as walking and cycling. Rapidly growing personal vehicles, poor quality, and insufficient public transport have resulted in inequality in service provision, due to which the urban poor are subjected to extreme hardship for their travel needs. As the limited road space without a functional footpath needs to be shared among pedestrians and slow- and fast-moving vehicles alike, urban transportation is faced with a large number of accidents involving urban poor and NMT users, congestion, poor energy efficiency, severe air and noise pollution, and carbon emissions leading to climate change. It is rather a quick journey from a "sustainable and environmentally friendly mobility" system to a "chaotic urban transportation" that is unsustainable and also inequitable.

Since the 11th Five Year Plan, Indian cities are competing with each other in developing metro systems in their respective cities, and quite a few of them would soon be flashing their metro smart cards as the new face of their transport system.

However, the marginalization of NMT continues to be evident from the metro development plans that they share with the public.

This necessitates a change in the way transportation is planned for future years. The city policy makers are in a dilemma on whether to focus on the provision of good-quality travel service or aim to control local and noise pollution or target the low carbon pathways to control carbon emissions. The present book has evolved from this "policy makers' dilemma."

Started off with the regional research initiative on energy, environment, and climate change by Sida (Swedish International Development Cooperation Agency) about 15 years ago, our research work on the transportation sector has been long and covered various important aspects including greenhouse gas mitigation; evaluation of alternative and cleaner options in transportation; barrier analysis; identification of appropriate policy measures to implement the alternatives; means to promote NMT; prioritization of alternatives, barriers, and policy measures using multi-criteria framework and AHP; integration of environmental concerns into long-term transportation planning; and incremental cost analysis to find the carbon competitiveness of the sector and sustainable mobility and its link to sustainable development goals (SDGs). These works, published by various agencies and journals, form the core material for this book. They have been cited at appropriate places along the text of this book. Although we are benefitted from the support of various international agencies such as Sida, the United Nations Development Program (UNDP), and the Institute for Global Environmental Strategies (IGES), Japan, at various stages in conducting the research, the opinions contained in this book are those of the author and should not be attributed to any of these agencies or its associates.

The material presented in this book has been developed over 15 years of research. The research work has evolved over time and followed closely the transformation of urban transportation. While this could address the issues as they evolved, some of the earlier works dated early 2000 may appear outdated for the sake of numbers. In the case of simple statistics, they have been updated with the most recent available information. However, if they represent an analysis aiming at comparative presentation of alternatives/scenarios, they have been left intact as the old dated information would not make any difference in the interpretation.

This book addresses the problems of urban transportation comprehensively by identifying potential alternatives, barriers for their implementation, policies for the implementation, and comparative analysis of different approaches of planning such as local pollution mitigation approach and global emission control approach. It considers the environmental dimension of sustainability and proposes a new and more effective policy framework. Such a framework would be of immense help to city and transportation planners and policy makers.

This book, in its attempt to develop a policy framework that is more effective, presents three different methodologies for prioritizing alternatives (barriers and policy measures), viz., analytic hierarchy process (AHP), multi-criteria group decision making with multi-stakeholder consultative process, and multi-criteria analysis with supra user and differentiated weights based on potential contribution toward the objective. Case studies are included for these three approaches which could help the reader in comprehending the methodology.

Starting with the basic planning and the need to look at sustainability principles (Chap. 1), this book goes on exploring the trends in urban transportation (Chap. 2). Difficulties in service provision owing to the changing urban dynamics are analyzed, and the environmental and climate implications of urban transportation are presented in Chap. 3. Chapter 4 presents all alternative options for environmental inclusion in transport and proposes a methodology for a comprehensive assessment of the same by means of energy-saving potential, emission reduction potential, and economic viability on life cycle operation. Chapter 5 identifies the problems in promoting NMT which can make commendable contributions toward environmental sustainability and social equity. Chapter 6 identifies barriers for NMT implementation and other cleaner alternatives such as clean fuels and alternative technologies. Such barriers have been prioritized using the popular analytic hierarchy process (AHP) so that the policy maker can easily pick a set of barriers that need to be addressed.

Chapter 7 identifies policy measures that are essential for the removal of barriers for the implementation of NMT and other cleaner alternatives in transport. These policy measures have been prioritized using two distinct methods of multi-criteria analysis. Chapter 8 presents all the regulatory and economic instruments that can be applied for the sustainability of urban transportation. Such a presentation is further augmented by various examples from all over the world. Means to integrate environmental concerns into urban transportation planning are attempted by developing a dynamic optimization model with environmental and carbon emission constraints. Chapter 9 applies the optimization model to analyze the co-benefits and carbon abatement costs of local emission control strategies and global emission control strategies.

This book aims to help the policy makers by suggesting and demonstrating a new policy framework to integrate sustainability and climate change mitigation into urban planning.

Mumbai, India

Sudhakar Yedla

Acknowledgments

My involvement with research on urban transport has been for 15 years, and I am happy to be consolidating it into a meaningful book. The material presented in this book is drawn from the research work carried out well over a decade, and I thank all my associates and colleagues who have helped me in various assignments and at different times.

I am grateful to all those who supported my research financially and technically by means of reviews and constructive comments at different platforms. Particularly, I wish to thank the Swedish International Development Cooperation Agency (Sida), Asian Institute of Technology (AIT), United Nations Development Programme (UNDP), and Institute for Global Environmental Strategies (IGES), Japan, for their generous support in conducting my research works on the transport sector.

Support by the parent institute, Indira Gandhi Institute of Development Research (IGIDR), has been remarkable, and the pleasure of working at IGIDR is the key driver behind this book. I wish to express my gratitude to the institute, the director, and all the former directors of IGIDR.

This book would not have been possible without the continued support of my family. Thank you everyone for inspiring me. Thanks to my son Dhiren and wife Namratha for their understanding and support, without which I would not have been able to complete this task.

Sudhakar Yedla

Contents

About the Author

Sudhakar Yedla is a professor of Environmental Policy at the Indira Gandhi Institute of Development Research (IGIDR), Mumbai, India. He is also a member of court (board) at the Central University of Jharkhand and guest faculty at the Indian Institute of Forest Management (IIFM), Bhopal. He has a PhD in Environmental Engineering from the prestigious Indian Institute of Technology (IIT) Kharagpur, India. He has a diploma in International Environmental Law from the UN Institute for Training and Research (UNITAR), Switzerland, and certification in Environmental Economics and Development Policy from the World Bank Institute, Washington, D.C., USA. His area of expertise includes urban environmental policy, climate change mitigation, and sustainable mobility. Earlier, he was a faculty member at the University of Ulsan, South Korea; Asian Institute of Technology, Thailand; and National Institute of Technology (NIT), Warangal. He had served the Institute of Global Environmental Strategies (IGES), Japan, as a policy researcher and the National Institute of Environmental Studies (NIES), Japan, as a fellow. He had also served UNDP as GEF and SDG consultant; UNESCAP as sustainable city consultant; and UNEP, UN-FAO, and Swedish International Development Cooperation Agency (Sida) as a project director for their regional initiatives on the roles of agriculture, climate change, and transportation. Prof. Yedla was also a visiting research fellow at the Korean Institute of International Economic Policy (KIEP), Seoul, South Korea. He has published extensively (over 100) in international journals, authored books, and served editorial boards of a number of international journals. He has been invited extensively by national and international organizations and universities to give lectures. He is a member of a large number of academic societies and won more than ten prestigious fellowships internationally. He has also been serving various corporate giants such as Cisco as a CSR reviewer.

Abbreviations

ASIF	Activity, structure, intensity, and fuel framework
BEST	Brihanmumbai Electric Supply and Transport
BIS	Bureau of Indian Standards
BOV	Battery-operated vehicle
BRTS	Bus rapid transit system
CC	Climate change
CERP	Common emission reduction potential
CNG	Compressed natural gas
CO	Carbon monoxide
CO_2	Carbon dioxide
CPCB	Central Pollution Control Board
CRRI	Central Road Research Institute
dB	Decibels
DHDS	Diesel hydro de-sulfurization
ERP	Emission reduction potential
ESP	Energy-saving potential
GDP	Gross domestic product
GHG	Greenhouse gas
GoI	Government of India
HC	Hydrocarbons
HEV	Hybrid electric vehicle
IGIDR	Indira Gandhi Institute of Development Research
JNNURM	Jawaharlal Nehru National Urban Renewal Mission
JPOI	Johannesburg Plan of Implementation I
JRY	Jawahar Rojgar Yojana
LCC	Life cycle operating cost
LNG	Liquefied natural gas
LPG	Liquefied petroleum gas
MAC	Marginal abatement cost

MMRDA	Mumbai Metropolitan Region Development Authority
MoEF	Ministry of Environment and Forests
MoUD	Ministry of Urban Development
MRTS	Mass rapid transit system
MUIP	Mumbai urban infrastructure project
MUTP	Mumbai urban transportation project
NAMP	National Air Quality Monitoring Programme
NCR	National Capital Region
NCT	National Capital Territory
NGV	Natural gas vehicles
NMT	Non-motorized transport
NMV	Non-motorized vehicle
NO_x	Oxides of nitrogen
NUTP	National urban transport policy
PAC	Pollution abatement cost
PAH	Polycyclic aromatic hydrocarbons
PJ	Penta joules
PKM	Passenger kilometer
PMGSY	Pradhan Mantri Gramin Sadak Yojana
PUC	Pollution under control
RSPM	Respirable suspended particulate matter
SO_x	Oxides of sulfur
SPCB	State pollution control boards
SPM	Suspended particulate matter
TERI	The Energy and Resources Institute
TKM	Ton-kilometer
TSP	Total suspended particulate matter
UN	United Nations
UNFCCC	United Nations Framework Convention on Climate Change
UNWCED	United Nations World Commission on Environment and Development
VAT	Value-added tax
VED	Vehicle excise duty
WHO	World Health Organization
WRI	World Resources Institute

List of Boxes

List of Figures

List of Tables

Chapter 1
Transportation Planning: The Basic Constructs

Economic development is the backbone of the societal progression, and transportation acts as the veins of the system. Connectivity for faster movement of goods and people and access to various activities are the key for the progress of a society. These are even more important in fast-growing urban centers. Due to the increasing migration and emphatic economic activity, metropolitan cities have been experiencing sharp rise in travel needs. At this juncture, where the rising travel demands are getting transformed into chaotic traffic scenarios and contributing significantly to local as well global emissions, it is important to plan the future expansions with well-informed decisions. This book is an attempt to present such issues that need to be considered and the process that needs to be employed while planning the urban transportation for the years to come. It presents a changed policy framework for such inclusive planning toward sustainable transportation in urban areas.

1.1 Basic Principles of Planning

Planning facilitates theoretical goals into actions. Good planning always aims for optimal solutions from the actions that are designed based on theoretical goals. Any such process aiming at an inclusive and sustainable planning should have the following characters (Litman 2013):

1. Comprehensive
2. Efficient
3. Inclusive
4. Informative
5. Integrated
6. Logical
7. Transparent

S. Yedla, *Urban Transportation and the Environment*,
DOI 10.1007/978-81-322-2313-9_1

It is important to consider all options and their impacts while planning a system and should aim to minimize the time and resource requirement. In order to ensure inclusiveness, the planning should involve, directly or indirectly, all the stakeholders such as users, citizens/tax payers, impacted residents, business entities, employees/workers, public officials, policy makers, affected organizations/interest groups, and legal experts. Providing all stakeholders with comprehensive information and understanding of the planning outcomes is important for the successful execution of the plan. It is important to have a logical connection between the components of the plan, both long and short term, and the transparency of the plan makes it more successful in reaching its goal.

Planning, transportation planning in particular, should also satisfy a set of resource constraints. In the modern planning, it is also important to consider sustainability elements both on the environment and social dimensions. Sustainability planning, unlike the conventional planning, strives for development rather than growth and considers all the recognized resource constraints and ecological risks and climate change impacts. The next section attempts to explain such elements of planning in transportation system.

1.2 Transportation Planning: Principles of Access, Affordability, and Sustainability

Planning of any transport system goes with an objective of achieving access with equitability and is subjected to affordability and a set of sustainability constraints. Factors that are considered essential for any such planning exercise include incrementality, accessibility, travel demand, alternative modes, equity principle, economic viability, land use, and sustainability principles. Each of these factors considered important for planning of transport system is explained below (Litman 2013; USDT 2010):

1.2.1 Baseline and Incrementality

Changes that a plan brings to the system are incremental to the baseline. Therefore, it is important to clearly define the baseline, taking into account trends that may affect future conditions such as population and economic growth. Such incrementality should also consider comprehensively all the negative externalities resulting from such planning so that they can be addressed in a complementary planning.

1.2.2 Accessibility

Accessibility is the basic construct of transportation planning and an ultimate goal of most transport systems. Accessibility refers to the ability to reach desired goods, services, activities, and destinations. For instance, walking, cycling, and public transit provide access to jobs, services, and other activities. Important factors that affect accessibility in any transport system are mobility, transportation system connectivity, land use, and mobility substitutes (Litman 2013).

Conventional planning tends to focus on transportation primarily in terms of mobility of vehicles, not the mobility of people and that jeopardizes the accessibility. Wide roads and expressways may improve the movement of vehicles and handle high traffic volumes but fail to facilitate the users of non-motorized modes of transport (NMT) users, which cater for a large share of work trips in cities. Such skewed transportation planning, coupled with city decongestion measures, can result in urban sprawl. Urban sprawl has a potential to make the transport system less efficient, particularly for urban poor and for the environment and energy consumption. More inclusive transportation planning aims at providing accessibility on all sections of the society on equitable terms.

1.2.3 Travel Demand

Planning transportation for the future needs information on the demand for travel. Travel demand refers to the amount and type of travel people need in order to meet the basic (work, education, etc.), essential (health care, firefighting, etc.), and optional (shopping, recreational, etc.) travel needs. The travel needs and behavior are affected by factors such as demographic (population, age, gender, lifestyle preferences), economic (jobs, income levels, activity, goods mobility), prices (fuel price, various taxes, tolls, parking fee, and insurance fees), transportation modes/option (NMT, public transport, carpooling, personal modes, facility of telepresence, remote office work), service quality (comfort, accidents, easiness in parking, congestion), and land use (density, vehicle mix on road, walkability, and connectivity). Any variation in these factors can potentially affect travel behavior. Future planning of transport system should ensure that all the travel demands are projected with due consideration to all the above factors and meet all the travel needs equitably.

1.2.4 Alternative Modes of Transport

Traditional transportation planning considers more of automobile modes of transport, which run on traditional fuels. However, NMT contributes significantly to work trips and hence needs to be augmented in the future planning of transportation (Yedla 2007). Therefore, the planning process should consider all modes of transport proportionate to their contribution to the travel services of various travel needs.

Due to various compelling factors, the fuel type and automobile technology have evolved, and a number of alternatives are available in both domestic and international markets. Such alternatives include unleaded gasoline, ultra-low sulfur diesel, compressed natural gas (CNG), biodiesel, hybrid technology, battery-operated vehicles (BOV), etc. Any future transport planning should consider all such alternatives available in order to have an inclusive and sustainable transportation system (Yedla 2004).

Such alternative modes are also driven by environmental and climate concerns. Choice of these alternatives is driven by their emission control penitential. While the cities in a developing country like India and China would not consider climate change mitigation strategies as part of their transportation planning, it is becoming increasingly evident that integrated approach toward local pollution control and climate change mitigation results in improved social well-being (Yedla 2007). Such an approach has been considered, and Chap. 9 presents a case study of *Mumbai*, to establish such a synergy.

1.2.5 Equity Principle

One of the most important aspects of transportation planning is to ensure equity in service provision. Such equity is represented as horizontal equity and vertical equity. While horizontal equity emphasizes the provision of service equally to all sections of the society based on economic status, vertical equity emphasizes prioritizing service provision to certain underprivileged and need-based sections of the society such as the physically handicapped, aged, and persons with certain disabilities. Transport planning has to ensure the equity in transport service provision.

Another dimension of equity is the service provision to urban poor which is a challenge in itself. Urban poor are at a disadvantage, as they need to travel long distances from fringe areas to the city center spending a major share of their monthly income. It is important to include transportation modes and land-use pattern that are suitable for urban poor in order to have an equitable transport system that caters the needs of all sections of the society.

1.2.6 Land Use

Land use planning designates the use pattern of the land within the city/metropolitan area limits. Such land uses include residential, commercial, institutional, industrial, transportation facilities, urban parks, brown fields, parks/urban greenery, agriculture, and forests/marsh lands (MMRDA 2011). It is important to have sufficient land area for the construction of roads and rail and other components of transport system. The adequacy in transport service provision is measured among other indicators by the share of land use for transport services.

For the city development to be sustainable, it is important to have efficient land-use pattern. Such an effective land-use pattern is evaluated based on various attributes such as density, mix, clustering, connectivity, built-up area, green space, accessibility, and facilitation of NMT (Litman 2013). A well-planned city should have land-use pattern satisfying all the attributes as listed above.

Land use planning sets the pathways to sustainability. For instance, land-use pattern decides the length of travel and time required for people to access their work and other needs. Land use planning can potentially control environmental emissions and greenhouse gas (GHG) emissions while reducing congestion. Land use planning can be used to promote walking to work and use NMT for other travel needs by locating the work places and residential complexes in close proximity. It also facilitates separation of emissions from transport and other sources so that the compounded effects can be minimized.

1.2.7 Economic Valuation

Alternative transportation options have different potential to address various issues of mobility and connectedness for different sections of the society. However, the overall transportation planning for any city is limited by the financial resources earmarked for this sector. Therefore, it is important to conduct economic viability study for all the potential alternatives before considering them for inclusion. As various options have different potential for energy saving, emission reduction, and climate change mitigation and also different economic viability, it is appropriate to consider life cycle operating cost (LCC) approach to have an integrated assessment of their potential as alternative transport options. An optimal mix of existing and alternative options should be identified, and such mix should be subjected to all the necessary limiting constraints of demand and resources. Chapters 7 and 9 demonstrate such an exercise in urban transportation in the form of a case study of *Mumbai* and *Delhi*.

1.2.8 Sustainability Principle

The transport sector is one of the major contributors to air pollution and greenhouse gas emissions (MoEF 2012). Hence, it has been attracting attention at all international deliberations on environment and climate. Accordingly the concerns on the transport sector and its impacts on energy and environment are reflected in Agenda 21 of the UN. With the overall objectives of reducing local and global emissions from all modes of transport sector, it seeks integration of environmental concerns into development of transport network. Chapters 7 and 9 of Agenda 21 have identified the following key issues in transport sector:

- Promoting integrated transport policies that consider alternative approached to meeting commercial and private mobility needs
- Integrating land use and urban and rural transport planning, taking into account the need to protect ecosystems
- Improving efficiency of transport and related sectors
- Accelerating phasing out of the use of leaded gasoline
- Promoting voluntary guidelines for environmentally friendly transport and action for reducing vehicular emissions
- Fostering partnerships at the national level for strengthening transport infrastructure and developing innovative mass transport schemes.

The Johannesburg Plan of Implementation (JPOI) further reiterates Agenda 21 goals for the transport sector and focuses on energy efficiency, access, integration, public transport, health, pollution, and emission issues. It further emphasizes the need to improve rural transport infrastructure in order to provide equal opportunities to the rural population.

The Rio+20 outcome document (UN 2012) emphasizes on achieving efficient movement of people and goods and access to environmentally sound, safe, and affordable transportation; developing energy-efficient and multi-model transport systems, efficient public transport, clean vehicles, and transport system; improving the transportation system in rural areas and their connectivity; and promoting integrated approach to policy making at the national, regional, and local levels for transport service and systems to promote sustainable development.

Sustainable development as defined by the United Nations World Commission on Environment and Development (UNWCED) is the development that "meets the needs of the present without compromising on the abilities of future generations to meet their own needs." A development path can be called sustainable when the interaction among the three spheres of sustainable development, viz., economy, social, and environment, is in a way getting the net positive benefits. While the interaction between society and environment deals with "environmental equity," interaction between economy and environment deals with "long-term resource allocation," and the interaction between economy and society deals with the "distributional efficiencies." This approach attempts to improve economic and social quality of life while limiting impacts on the environment to the carrying capacity/assimilation capacity of nature. In such a system, the net benefits of interactions are positive (Yedla 2008). Any predicted imbalance in the system in the form of water and energy supply limits, climate change, etc. would adversely affect the quality of human life and the economy. Sustainable development approaches to development essentially addresses these issues.

Applying such principles to the transportation sector, a widely accepted definition of a sustainable transportation system has evolved. A sustainable transportation system is one that (AASHTO 2009):

- Allows the basic access needs of individuals and societies to be met safely and in a manner consistent with human and ecosystem health and with equity within and between generations
- Is affordable, operates efficiently, offers choice of transport mode, and supports a vibrant economy
- Limits emissions and waste within the planet's ability to absorb them, minimizes consumption of nonrenewable resources to the sustainable yield level, reuses and recycles its components, and minimizes the use of land and the production of noise

Any transportation system developed on such sustainability principles augments the economic development with least impacts on environment and also adds to the good quality social living. Planning transportation and delivering solutions and infrastructure in a more sustainable manner must ensure equity in the following domains:

Economy: Ensures to support economic vitality while developing infrastructure in a cost-efficient manner. Costs of infrastructure must be within a society's ability and willingness to pay from all strata of the society. User costs, including private costs, need to be within the ability of people and households to pay for success. No additional burden should be levied on the disadvantaged sections of the society.

Social: Ensures to meet social needs by making transportation accessible, safe, and secure and includes provision of mobility choices for all people (including people with economic and physical disadvantages)

Environment: Creates solutions that are compatible with and that can be an enhancement to the natural environment, reduce emissions and pollution from the transportation system, and reduce the material resources required to support transportation

1.3 City and Transportation Planning Models

Urban planning is classified into models based on the land-use planning philosophy. One model is based on *urban sprawl*, and the other one on *compact growth* principles. While the sprawl refers to dispersed, low-density, automobile-dependent land use, the smart growth refers to more compact, mixed, multi-modal land use (Litman 2014). These two are distinct models having contrasting characteristics. Sprawl as it is developed outwards and has large space availability can potentially facilitate in having maximum roadway capacity and speed, more parking space, and low road user charges. However, it results in high public transportation charges (as the distance of travel increases), inferior public transport, and poor conditions for walking and other NMT use. In contrary to this, the smart growth model facilitates improvements in transit services, cheaper public transport that is affordable to all sections, and improved conditions for NMT. The smart growth or compact city model reduces parking supply, limits road space, and restricts traffic speed.

Conventional transport planning focuses more on sprawl, particularly in Asian region. Sprawl is linked to automobiles, and during the past few decades, the cities in Asia have experienced urban sprawl and increased number of vehicles. The land use also followed the pattern of a sprawl. Sprawl is criticized for being environment and climate unfriendly and energy intensive. As a means to bring in sustainability in transportation along with other sectors, smart city or compact city model has been developed and also attempted in some cities in China, Japan, and also in India. Compact city model, due to its closed geographic area, reduces the need to travel and promotes public transport and discourages the use of personal modes of transport. It promotes the use of NMT making it one of the best city growth models for environment and natural resources.

While the compact city principles can be applied to newly developed habitats and cities, the existing cities which are largely characterized by urban sprawl model of growth need to be augmented for the sustainability. This book, from Chaps. 2, 3, 4, 5, 6, 7, 8, and 9, attempts to develop a new policy framework by incorporating certain elements of sustainability into transportation planning.

References

AASHTO (2009) Transportation and sustainability best practices background. Gallaudet University Kellogg Center, Canada

http.www.indiastat.com (viewed on 1 Aug 2014)

Litman T (2013) Planning principles and practices. Victoria Transport Policy Institutions, Victoria

Litman T (2014) Evaluating criticism of smart growth. Victoria Transport Policy Institutions, Victoria

MMRDA (Mumbai Metropolitan Regional Development Authority) (2011) Regional plan for Mumbai metropolitan development area. MMRDA, Mumbai

MoEF (Ministry of Environment and Forests) (2012) India – second communication to United Nations Framework Convention on Climate Change. Ministry of Environment and Forests, Government of India, New Delhi

United Nations (2012) The future we want – Rio+20. The United Nations Conference on sustainable development, A/Conf.216/L.1*

USDT (US Department of Transportation) (2010) The transportation planning process: key issues – a briefing book for transportation decision makers, officials, and staff. Federal Highway Administration, Washington, DC

Yedla S (2004) Transportation trends, alternatives and policy issues. In: Radhakrishna, Parikh (eds) India development report 2004–05. Oxford University Press, New Delhi

Yedla S (2007) Non-motorized modes of transportation for sustainable mobility: strategies for its adaptation in Mumbai, India. In: Air pollution control in the transportation sector: third phase research report of urban environmental management project. Institute for Global Environmental Strategies, Japan, pp 21–82

Yedla S (2008) Scoping paper for the establishment of a UNESCAP eco-city development award. A report submitted to ESD Division of UN-ESCAP, Bangkok

Chapter 2
Urban Transportation Trends: An Overview

2.1 Introduction

One of the major concerns of urbanization is the sprawling which directly and seriously affects the transportation system. With the growing urbanization, Indian cities have been facing outrageous growth in travel needs and the vehicle stock. The number of motorized vehicles in Indian cities has outgrown the provision of transport infrastructure. This trend is predominant in metropolitan cities owing to their higher economic growth rates and increasing rate of urbanization. Road infrastructure has not developed commensurately to support high growth in vehicle stock. This has resulted in increasing congestion, air and noise pollution, longer travel time and distance, and more accidents. Poor infrastructure results in poor fuel efficiency as well. Increased dependence on personalized modes of travel is apparent from their rising stock in metropolitan cities. Various policy initiatives taken by the Government of India have succeeded to some extent in controlling urban air pollution in some cities; however, the ever-rising vehicular stocks and lag in the application of economic tools to control traffic growth and the resulting pollution keep the situation at the same level as before. Delhi, the leading metropolitan city in India, is among the most polluted cities in the world. It is experiencing very high levels of air pollution, mostly coming from urban transportation. Though air quality in Delhi has improved after the implementation of CNG, it has started to decline soon after. Many class II cities like Hyderabad, Kanpur, and Varanasi are closely following the trends similar to what Delhi followed in the early 1990s.

This chapter presents the trends in urban transportation development and various national initiatives implemented to control the growth in transportation and the sectors' environmental emissions. The chapter discusses fuel alternatives, management measures, and policy measures taken up to achieve sustainable transportation in Indian urban centers.

Part of this chapter are derived from Yedla 2004.

S. Yedla, *Urban Transportation and the Environment*,
DOI 10.1007/978-81-322-2313-9_2

2.2 Issues in Urban Transportation

Efficient system of transportation makes key contributions to the economic growth, competitiveness, and social inclusion. Addressing the issue of urban transportation is a complex exercise, and any efforts to achieve sustainable transportation need to be holistic covering diverse aspects of travel demand, vehicular growth patterns, emissions, auto fuels, traffic management, and efficient land-use pattern. Attempts to arrive at a solution from any of these dimensions in isolation would not achieve sustainability as most of these are essentially interlinked and need to be addressed in integration.

2.2.1 Growth in Vehicular Stock

All million plus cities in India are facing consistent rise in vehicular stock and growth in demand for travel. Among all metropolitan cities, Chennai, Delhi, Kolkata, and Mumbai have registered the highest growth rate of vehicular stock. Delhi stands out among all cities with a total vehicular population equal to the aggregate vehicular stock of all the other three metropolitan cities, viz., Chennai, Kolkata, and Mumbai. Rapid expansion of city boundaries and increased number of suburbs to cater for migrating populations could be the reasons for such high growth. Table 2.1 presents the growth rates of vehicular stock in all major cities in India from 1985 to 2005. Class I cities are competing closely with the five megacities on the annual growth rates of vehicular stock. Cities like Surat, Pune, Jaipur, and Nasik have experienced tremendous growth in vehicular stock over the last two decades and are trying to catch up with the megacities. Hyderabad, Ahmedabad, and Bangalore are already comparable to the megacities with respect to the vehicular stock. Given the fact that unlike other megacities, no concrete measures/efforts are on to check the travel growth or pollution in these cities, very soon they would reach alarming levels of traffic and resulting environmental emission. Figure 2.1 presents the change of vehicular stock in different cities from 2001 to 2011.

Among the total vehicular stock, personalized modes of transport are dominating in almost all cities. Table 2.2 presents the mode-wise breakup of vehicular stock in different cities for the year 2012. Delhi and Mumbai roads are predominantly occupied by two-wheelers, and in the case of Kolkata, it is cars showing dominance. Lack of proper/efficient public transportation in Delhi could be the reason for rapid growth of two-wheeler population over time. However, it is an interesting observation that their growth rate has started declining in the very recent times. In Delhi, where two-wheelers were registering at the rate of around 0.1 million per annum, the registration has come down to around 50,000 per annum by 2002. Considering that old two-wheelers above certain age get phased out, in Delhi the total number of two-wheelers plying may actually be reducing (MoPNG 2002). In contrary, the

Table 2.1 Percentage annual growth rates of motor vehicles in major Indian cities from 1985 to 2005

City/district	1985–1990	1990–1995	1995–2005	Population in millions (2005)
Greater Mumbai UA	9.05	5.87	11.14	18.78
Kolkata UA	19.44	24.89	9.53	14.22
Delhi state	18.93	9.71	6.57	15.02
Chennai UA	29.27	11.05	10.49	6.96
Bangalore District	17.84	9.11	14.43	6.5
Hyderabad UA	26.96	13.66	9.76	6.46
Ahmadabad District	16.39	10.76	10.76	5.07
Pune District	14.35	14.13	11.95	4.41
Surat District	20.87	9.18	11.42	3.63
Jaipur District	17.24	11.97	10.79	2.73
Lucknow District	5.39	5.48	6.93	2.67
Nagpur District	24.4	8.94	11.22	2.34
Patna District	20.77	10.23	13.51	1.98
Indore District	16.98	10.14	12.95	1.88
Vadodara District	26.37	14.28	13.83	1.72
Bhopal District	6.02	8.84	5.96	1.69
Ludhiana District	23.71	9.52	15.95	1.54
Visakhapatnam District	13.66	14.32	15.32	1.58
Varanasi District	16.72	10.33	9.96	1.3
Nashik District	23.46	10.38	11.75	1.2
Jabalpur District	21.31	8.25	14.05	1.18

Source: Compiled by authors using MoSPI (2013)

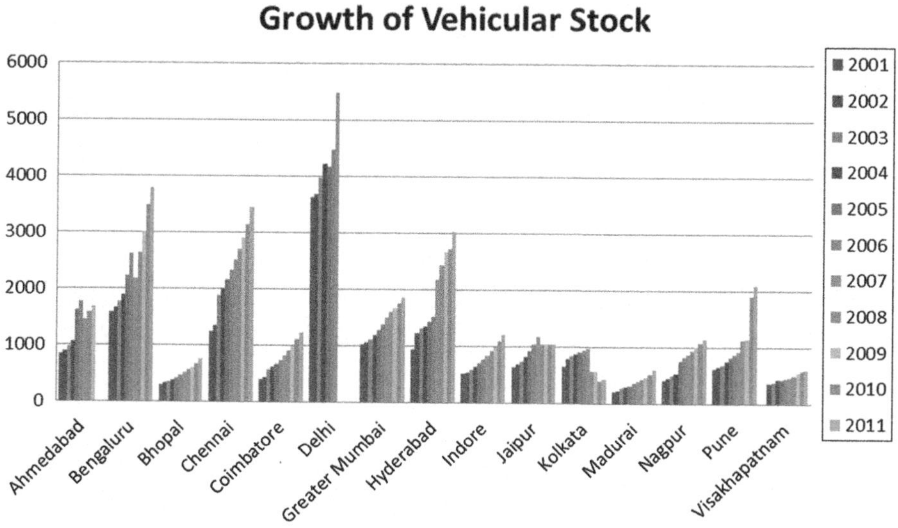

Fig. 2.1 Growth of vehicular stock in different cities from 2001 to 2011 (Source: Developed by the authors using data from http.www.indiastat.com)

Table 2.2 Total registered motor vehicles in major cities of India in the year 2012

	Two-wheelers	Cars and taxis	Jeeps and omnibuses	Tractors and trailers	Buses	LCV	HCV	Others	Total
Agra	515,154	55,021	6,322	35,736	10,515	14,294	3,866	1,654	642,562
Allahabad	584,248	57,003	7,671	8,746	10,296	8,010	8,929	2,611	687,514
Aurangabad	193,878	13,339	7,302	10,786	12,768	11,733	3,572	1,022	254,400
Bengaluru	2,624,707	757,087	53,598	13,162	163,503	73,888	76,843	20,046	3,821,304
Bhopal	602,793	101,392	1,482	16,528	17,395	11,795	7,390	1,206	759,981
Chennai	2,398,366	681,181	21,127	14,253	142,495	82,457	97,996	40,743	3,478,618
Coimbatore	1,023,414	154,382	5,682	8,372	18,373	15,406	10,632	13,430	1,249,691
Delhi	4,395,086	2,178,442	168,856	5,483	88,795	124,547	4,792	519	6,966,520
Mumbai	1,044,829	616,674	27,835	843	122,453	53,969	8,160	1,512	1,876,275
Hyderabad	2,144,410	524,278	34,650	8,283	153,171	96,642	120,718	6,892	3,089,044
Indore	930,223	163,010	4,397	25,706	25,190	27,741	46,760	2,596	1,225,623
Jabalpur	462,632	40,582	2,081	14,919	17,475	11,236	10,637	3,655	563,217
Jaipur	1,248,076	231,824	49,668	44,352	47,475	14,116	75,359	55	1,710,925
Kanpur	849,098	98,580	10,166	6,630	7,446	2,102	18,135	3	992,160
Kochi	221,157	123,615	11,185	257	28,527	24,635	5,290	4,021	418,687
Kolkata	182,087	225,985	0	82	23,745	0	15,235	701	447,835
Lucknow	970,897	171,784	16,363	19,127	10,293	14,452	6,683	3,512	1,213,111
Madurai	493,575	52,311	1,627	7,605	23,289	9,983	12,070	7,683	608,143
Meerut	306,202	46,540	1,133	58,690	5,861	3,125	4,231	0	425,782
Nagpur	967,838	92,386	29,086	10,648	22,285	21,027	16,481	1,153	1,160,904
Nashik	300,877	36,195	8,176	18,241	18,202	12,348	5,215	757	400,011
Patna	448,104	95,286	0	22,974	43,027	0	28,776	4,200	642,367
Pune	1,551,968	297,139	41,142	27,770	61,798	53,002	24,877	4,462	2,062,158
Raipur	412,707	38,426	6,028	23,780	8,367	12,354	27,039	2,840	531,541
Varanasi	426,522	44,758	8,768	22,960	16,834	13,869	9,381	400	543,492
Vijayawada	340,614	40,323	2,935	11,853	23,025	2,398	3,397	724	425,269
Visakhapatnam	469,784	72,095	5,702	12,976	38,641	10,348	14,586	1,954	626,086

Source: Compiled by the author using data from http.www.indiastat.com

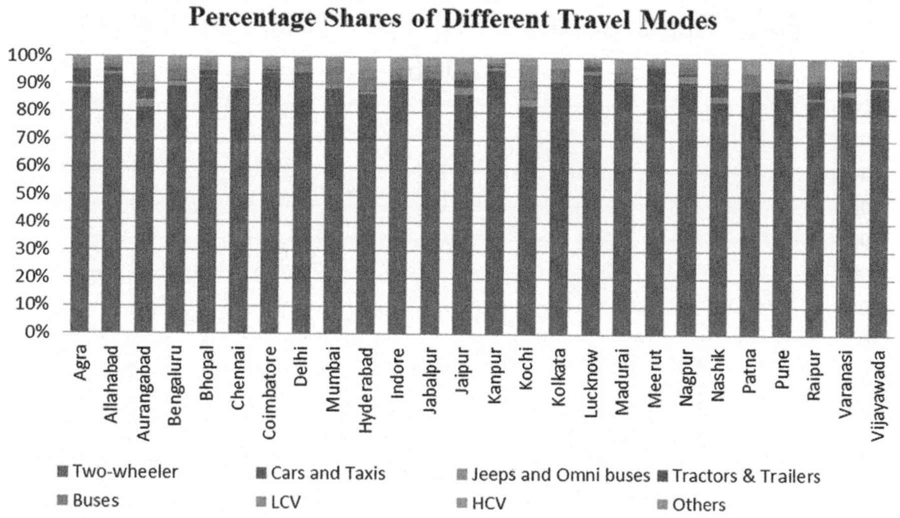

Fig. 2.2 Percentage shares of travel modes in different cities in the year 2012 (Source: Developed by the author using data from http.www.indiastat.com)

trends observed in the other cities are fast growing. The number of two-wheelers as a percentage of the total registered vehicles is considerably high among the small and industrialized cities like Pune, Surat, and Kanpur. Figure 2.2 presents the percentage share of each travel mode in various cities of India in the year 2012. City/public transport in small cities is fairly low compared to the megacities like Delhi and Mumbai. This could be due to the fact that the distance traveled to the workplace is considerably less in smaller cities compared to that of Delhi and Mumbai. Further, it could be attributed to the fact that these megacities have undergone a proper transportation planning and development of public transport so as to cater for impending travel needs, whereas the class II cities are still adapting add-on approach.

The share of public transport like buses and auto-rickshaws is high in megacities compared to the rest. Hyderabad shows alarming trends of growth in two-wheeler population, which is a dangerous indication of worsening ambient air quality.

2.2.2 *Vehicular Density*

Density of vehicles is presented as number of vehicles per thousand population. It indicates the vehicle ownership intensity of various cities/states in the country. For the entire country, the number of vehicles per 1,000 population is 109. Figure 2.3 presents the number of vehicles per 1,000 population for different states in India. Uttar Pradesh and Goa are the two states with very high vehicle density (vehicles per 1,000 population), while Bihar and West Bengal apart from the northeastern

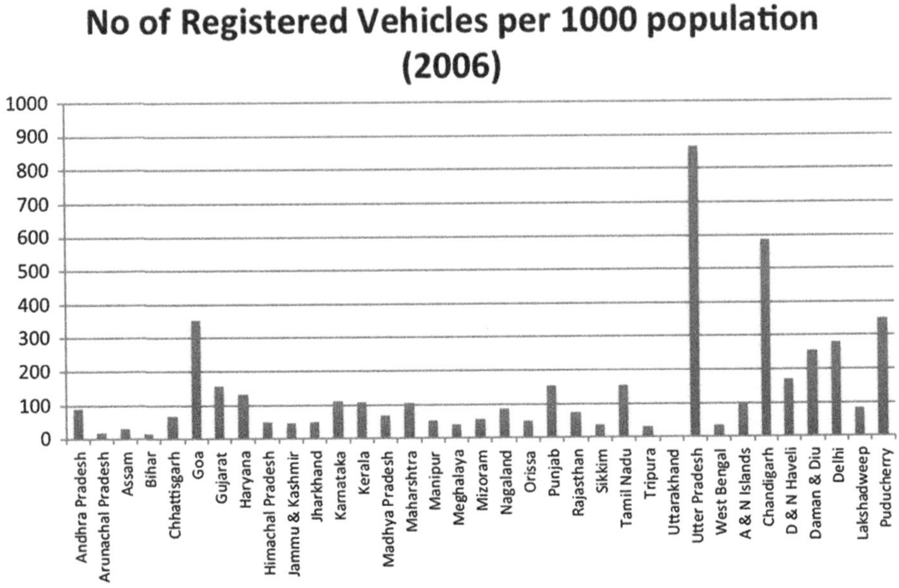

Fig. 2.3 Vehicle density (vehicles per 1,000 population) in different states (Source: Developed by the author using data from http.www.indiastat.com)

states are the lowest in vehicle density. Union territories except A & N Islands are very high on vehicle density. Increase in density results in road congestion particularly in urban centers.

2.2.3 *Travel Demand*

Different modes of vehicles have different occupancy rate. Hence, the effectiveness of any transportation mode is measured in terms of passenger kilometers (PKM)/ton kilometer (TKM) served by the respective mode. Two-wheelers have very poor occupancy (1.5 persons/vehicle) against buses (37 persons/vehicle). Passenger kilometers catered by buses are much higher than any other mode, and hence, they cover major share of the travel demand in many cities. Due to the high occupancy rate, emission per PKM is very low in the case of buses as compared to that of cars and two-wheelers. This makes the bus not only an effective transport mode but also environmentally efficient.

Owing to the rapid expansion of cities, PKM, an indicator of travel demand, is expected to grow by leaps and bounds in the major metropolitan cities in the years to come. In a study carried out by IGIDR employing econometrics and spreadsheet models, the total passenger travel demand in Delhi was estimated to increase from 73 PKM to 253 billion PKM during 1997–2020, registering an annual growth rate of 5.3 %. Freight travel demand is also expected to increase from 2.63 billion TKM to 7.18 billion TKM during the same period (IGIDR 2001). During the same period, total passenger travel demand for Mumbai is expected to grow from 32 billion to 137 billion PKM with an annual growth of 6.8 %. Flight travel demand is estimated to rise from 0.36 billion to 1.37 billion TKM. Therefore, transportation planning should aim not only to augment the infrastructure to support the increasing number of vehicles but also to improve public transport such as the metro rail system, monorail, and bus rapid transit system (BRTS).

Such systems are being implemented in all major metropolitan cities in India. In the absence of such mass transport means, the increasing travel demand can lead to more energy consumption and pollution generation. In spite of mass transport being promoted, personalized vehicle increases over time, and in order to minimize the pollution loads resulting from the increasing vehicle fleet and the use of poor-quality fuels, technological options will have to be explored in terms of improved automobile engines and fuels and also promotion of NMT.

2.2.4 *Road Density*

The number of vehicles per kilometer road length presents the infrastructure sufficiency. The average number of vehicles per kilometer road length for India is 31 (2011). Figure 2.4 presents the number of vehicles per 1,000 km road length for the different states in India. Uttar Pradesh has a very large number of vehicle

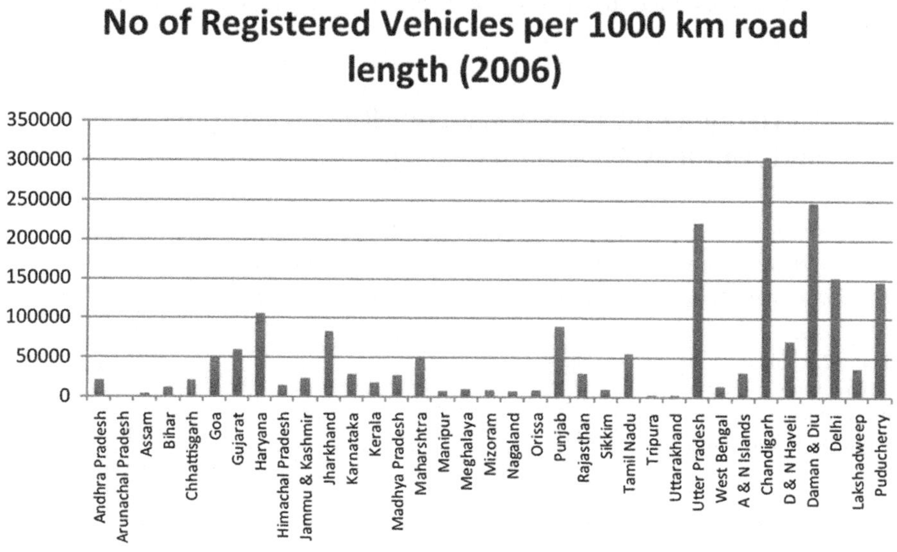

Fig. 2.4 Vehicle density (vehicles per 1,000 km road length) in different states (Source: Developed by the author using data from http.www.indiastat.com)

registers per 1,000 km road length in spite of the fact that Uttar Pradesh has a large road network. Punjab and Haryana follow UP on the number of vehicles per 1,000 km road length. All union territories exhibit high vehicle densities, which could be due to the twin effects of higher number of vehicles and low availability of road infrastructure. These indicators explain the congestion levels in Indian road transport sector.

The average density of roads (area of roads per 1,000 km^2 of land) in India is 965 km, and the per capita road density per 1,000 population (year 2008) is 2.77 km (excluding the roads under Jawahar Rojgar Yojana (JRY) and Pradhan Mantri Gram Sadak Yojana (PMGSY)). The road density in urban areas and rural areas is 3,893.48 and 920 km per 1,000 km^2, respectively. Road density per 1,000 population is 0.91 km and 3.55 km for urban and rural areas, respectively.

2.2.5 *Share of Different Modes in Providing Travel Service*

While the infrastructure development is more skewed toward motorized vehicles, NMT takes a major share of work trips in Mumbai and other cities in India. Table 2.3 presents the share of different modes used by the poor in their work trips. While 67 % of the poor uses NMT for their work trips, rail and bus cater for 30 % work trips. Urban poor are particularly dependent on NMT for various travel needs. Given the fact that it is used by the poor for their access to work and also the fact that it is the most energy efficient and environment and climate friendly, more emphasis should be given for the development of NMT in the future transportation planning. Chapters 5 and 6 discusses in detail the issues related to NMT.

2.3 Environmental Emissions from Road Transport and Ambient Air Quality

The old and poorly maintained vehicles and the growing vehicular stock result in increased environmental emission. The transport sector contributes a major share of environmental pollution (around 70 %). CO is the major pollutant coming from the

Table 2.3 Share of usage of different modes of transport by the poor in Greater Mumbai (in percentage)

Mode	Percentage share
NMT (foot and bicycle)	66.9
Rail	16.1
Public bus	14.5
Three-wheeler	1.3
Taxi	0
Two-wheeler	0.7
Car (own car+ other car)	0.2

Source: Compiled by the author from various sources

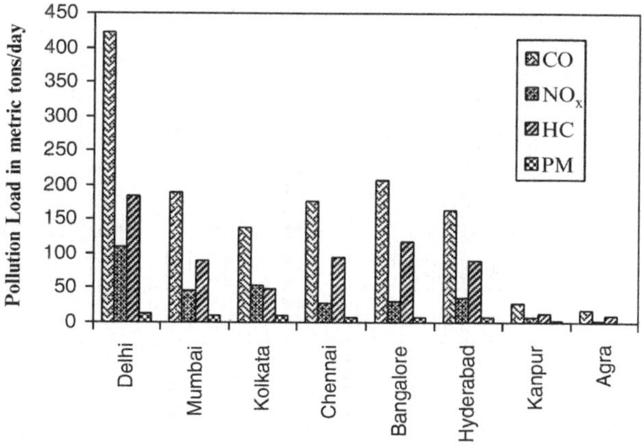

Fig. 2.5 Pollutant load from the transport sector in different cities in India (Source: Developed by the authors using data from CPCB 2012)

transport sector, contributing to almost 90 % of the total emission. Hydrocarbons follow closely. It is observed that the contribution of the transport sector to the particulate pollution is as less as 3–5 %. Most of the SPM is due to the resuspension. Pollutant emission levels have gone up substantially owning to the fast growth in vehicular stock. Delhi emits about 420 metric tons of CO everyday with almost 1,400 metric tons of total pollution every day. Daily emissions of various pollutants from the transport sector are presented in Fig. 2.5. Similar trends are observed in the other metropolitan cities such as Hyderabad, Bangalore, and Pune (Yedla 2004).

Apart from the increasing stock, it is also reported by many researchers that the growing trend of emissions is observed due to the fact that the vehicles are used for extended lifetime without proper maintenance. Ill-maintained vehicles tend to emit more pollutants than others. Improper inspection and maintenance, use of poor-quality fuels, poor road conditions, and increased congestion add to emissions. At present, owing to many initiatives from various sectors, most of the abovementioned factors are showing improvement.

2.3.1 Ambient Air Quality in Different Cities

The Central Pollution Control Board (CPCB) has been assigned various functions under the *Air Act* 1981 to plan nationwide programs for the prevention, control, and abatement of air pollutants. Accordingly, the *National Air Quality Monitoring Programme* (NAMP) had been initiated during 1984–1985 at a national level, which had been steadily expanded and at present comprises 290 monitoring stations covering 90 cities in 24 states and 5 union territories. The NAMP network is operated with the involvement of various agencies, viz., State Pollution Control Boards (SPCBs), CPCB headquarters and zonal offices, pollution control committees, and research institutes. The air quality parameters regularly monitored nationwide are SPM (8 hourly for 24 h), sulfur dioxide (SO_2), and nitrogen oxide (NO_x)

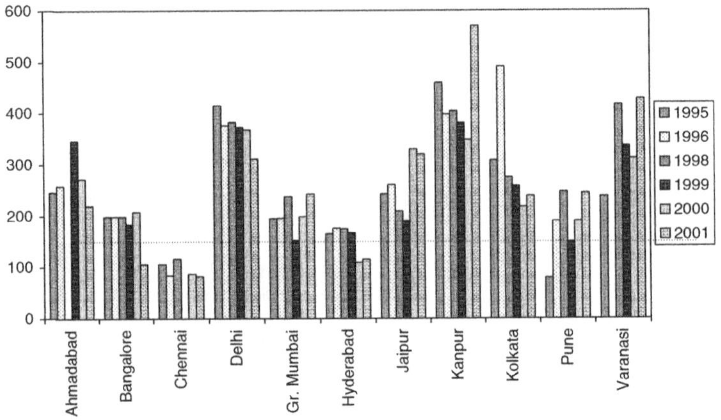

Fig. 2.6 Changes in SPM concentration (Source: Developed by the authors using data from CPCB 2012)

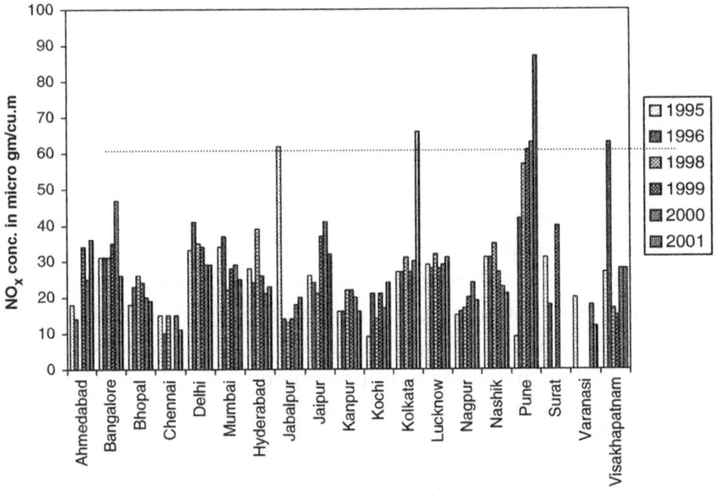

Fig. 2.7 Changes in NO_x concentration (Source: Compiled by the author using data from CPCB (2001) National Ambient Air Quality Monitoring Series (NAAQMS/8,10,15,21))

(4 hourly for 24 h), while additional parameters like carbon monoxide (CO), polycyclic aromatic hydrocarbons (PAH), ozone, respirable suspended particulate matter (RSPM), benzene, and trace metals are additionally monitored in Delhi and some other cities, but not in all the cities.

SPM is the most common air quality indicator, which exceeds the permissible level in many cities. SPM levels in different cities across the country from 1995 to 2001 are presented in Fig. 2.6. NO_x is another important air quality indicator whose dynamics are presented in Fig. 2.7. The variation in air quality is a mixed scenario

Table 2.4 Share of air pollution from different sectors

Pollutants	Sector contributing to pollution	Percentage contribution	
		Delhi	Mumbai
CO	Transport	76–90	92
	Industrial	37–13	8
	Domestic and others	10–16.3	0
NO_x	Transport	66–74	60
	Industrial	13–29	40
	Domestic and others	1–2	0
SO_2	Transport	5–12	2–4
	Industrial	84–95	82–98
	Domestic and others	0–4	0–16
PM	Transport	3–22	0–16
	Industrial	74–16	34–96
	Domestic and others	2–4	53–56

Adopted from the expert committee report on auto fuels (MoPNG 2002)

as some cities follow a rising trend with others experiencing an improvement. Kolkata, Bangalore, and Pune are experiencing a rise in NO_x over time. Delhi showed a steady fall after initial rise.

2.3.2 *Share of the Transport Sector*

Air pollution comes from various natural sources as well as anthropogenic sources. For certain pollutants like CO and HC, the major source has been anthropogenic, and others like SPM are contributed by natural sources. Even among anthropogenic sources of pollution, various sectors influence different pollutant levels in a different way. For instance, the transport sector contributes most of the CO emission. The industrial sector contributes most of SO_2. Table 2.4 presents the summary of share of various pollutants from different sectors. This depends on the location, activity, and prevailing meteorological conditions. As it can be observed from the table below, the share of sectors for various pollutants is different from Delhi to Mumbai.

2.3.3 *Vehicle Harboring and Pollution*

While the registered vehicles contribute to air pollution in urban centers, a major share of vehicles that are harbored outside the city center also contributes significantly to air pollution in urban areas. Most of the cities experience significant inflow of vehicles and passing of vehicles. In a study carried out by Central Road Research Institute (CRRI), Delhi, based on the fueling pattern and their garaging character, it was observed that a majority of commercial vehicles (trucks) filling fuel in the city fuel stations are garaged (housed) outside the city. This fact particularly influences the transportation planning and the emission estimation. Table 2.5 presents the details of garaging status by vehicle type in different cities. Most of the commercial vehicles have a major share garaging outside the city in almost all cities. It is interesting to observe that this fraction is almost two times in the case of Delhi. With the fact that commercial vehicles are more polluting and these are "unaccounted part" of the stock, Delhi ended up experiencing more pollution than the estimated levels. Among personal vehicles all cities have similar character except in the case of Hyderabad where the inflow of cars and two-wheelers is considerably high. This could be due to the fact that Hyderabad does not have suburbs. As a result people stay in neighboring places and commute to the city for daily needs. Hence, it is essential to account for this "nonresident" vehicle stock in the estimation of emission, infrastructure development, and also the transportation planning.

2.4 Alternatives Options in Urban Transportation

The following are the alternative fuels, technologies, and management measures attempted at improving transportation systems globally:

1. Natural gas vehicles (NGV) including liquefied petroleum gas (LPG)
2. Electric vehicles
3. Hybrid electric vehicles
4. Fuel cells
5. Hydrogen energy
6. Biofuels
7. Inspection and maintenance (I&M) and certification system
8. Retrofitting of emission control devices
9. Traffic management and construction of bypasses
10. Construction of metro rail, monorail, and BRTS

A detailed description and analysis of these alternatives is given in Chap. 4 of this book.

2.5 National Initiatives to Address Transportation Issues

Standards for controlling emission levels of new vehicles were incorporated into the Motor Vehicles Act (1989) as late as 1991, enforcing some quality control on the automobile industry. The mass emission standards refer to gram of pollutant

Table 2.5 Garaging status of vehicle type in different cities

Vehicle type	Garaging status (%)											
	Delhi		Mumbai		Kolkata		Chennai		Bangalore		Hyderabad	
	Inside	Outside	Inside	Outside	Inside	Outside	Inside	Outside	Inside	Outside	Inside	Outside
Cars	90.42	9.58	92.11	7.89	95.26	4.74	84.00	16.00	97.83	2.17	75.72	24.28
Two-wheelers	92.20	7.80	95.30	4.70	82.56	17.44	95.40	4.60	96.93	3.07	80.46	19.54
Autos	98.56	1.44	96.33	3.67	86.13	13.87	92.40	7.60	96.79	3.21	86.04	13.96
Buses	83.08	16.92	100.0	0.00	87.83	12.17	70.70	29.30	77.78	22.22	79.31	20.69
LCVs	73.53	26.47	94.46	5.54	83.65	16.35	66.00	34.00	83.76	16.24	53.53	46.47
HCVs	30.51	69.49	88.45	11.55	67.56	32.44	66.30	33.70	68.29	31.71	50.10	49.90
MCVs	35.62	64.38	70.31	29.69	66.66	33.34	71.70	28.30	50.00	50.00	65.85	34.15

Source: Compiled by the authors using data from different sources

emitted per kilometer traveled under specified driving conditions as notified by the Ministry of Surface Transport and enforced by the State Transport Department. Though the Bureau of Indian Standards (BIS) for fuel started incorporating emission parameters, the standards were set according to the terms and conditions dictated by the industry. Both the regulatory authorities and the manufacturers pleaded incapacity to make any drastic improvements, and so the standards they followed were far too lax compared to those elsewhere in the world. In 1996, the government came up with mass emission standards for vehicles, which are stricter than earlier. Other measures during 1994–1996 aimed at improving emissions from the transport sector are the use of unleaded petrol and fitting of catalytic converters in the car to reduce CO emissions. Subsequently EURO I, II, and III norms also have been enforced followed by India's own emission standards Bharat Stage (MoPNG 2002; Yedla 2004).

Poor maintenance of vehicles adversely affects their emission efficiency. The role of maintenance in combating vehicular pollution was reflected in government policy for the first time in 1989, which made the certificate of fitness as mandatory for registration of public vehicles, commercial vehicles, and personal vehicles older than 15 years. The 1990 vehicular emission rules required all motor vehicles to comply with the laid-down exhaust emission standards. The transport department issues Pollution Under Control (PUC) certificates to vehicles. Vehicles owners are required to check the emission level of their vehicles every 3 months and obtain a PUC certificate. Vehicles failing to meet the standards are required to rectify the fault and obtain the certificate. The State Transport Authority fines vehicles not possessing a PUC certificate.

This is a step toward minimizing vehicular pollution by regular checks. This system, however, has come under severe public criticism due to the existing lacunae in the issue of certificates and the discrepancies in pollution readings from one station to another. Though there is a provision to levy a monetary fine of Rs 1,000 on motorists who fail to abide by the law, the enforcement has been very poor. Statistics maintained by the Automobile Association reveals that more than 50 % of vehicles in Delhi in May 1995 failed to comply with the prescribed standards. It is even more alarming that nearly 44 % of the new vehicles checked were found to be not in compliance with the standards. This shows that PUC despite being a potentially powerful instrument in controlling pollution from vehicles has failed to make an impact on vehicular pollution. This, apart from cutting down the emissions, actually created a lot of awareness among citizens for the environmental pollution.

The failure of the administration to enforce environmental regulations has led to judicial interventions. The Supreme Court has come up with several guidelines in the last few years. The Court has urged the government to accept the emissions standards EURO I, II, III, IV, etc. for the vehicles as adopted by the European Commission. India has developed its own standards, namely, Bharat Stage I, II, II, IV, etc. In last few years, the Supreme Court of India and High Court of Maharashtra and other states have issued a number of directives aiming at emission control, which are listed in Boxes 2.1 and 2.2.

Box 2.1: Directives of the Supreme Court Aiming at Environmental Emission Control

- Elimination of leaded petrol from NCT of Delhi by September 1, 1998.
- Phasing out of all commercial vehicles which are more than 15 years old by October 2, 1998.
- From June 1, 1999, Euro I norm was made effective for all private vehicles.
- No 8-year-old buses to ply except on CNG or other clean fuels by April 1, 2000.
- Entire city bus fleet to be steadily converted to single fuel mode on CNG by March 31, 2001.
- Replacement of all pre-1990 automobiles and taxis with new vehicles on clean fuels by March 31, 2000.
- From April 1, 2000, no vehicles will be registered in the National Capital Region, unless it conforms to EURO II norms.
- Supply diesel with 0.05 % sulfur content in the NCR from May 2002.
- Supply petrol with 1 % benzene content in the NCR from October 31, 2001.
- Supply of only premix petrol in filling stations to two-stroke engines by December 31, 1998.
- Ban on registering two-stroke vehicles from July 2000.
- All in-use vehicles with two-stroke engines will have to be fitted with catalytic converter.

Source: CPCB (1999)

Box 2.2: Recommendations by the Committee Constituted by Mumbai High Court

- The sulfur content in the entire diesel to be supplied in Mumbai City at all the petrol pumps should be reduced to 0.05 % by October 1, 2000. It should be further reduced to 0.035 % by April 1, 2003, and to 0.005 % by April 1, 2005.
- The benzene content in all the petrol supplied in Mumbai City at all the petrol pumps should be reduced from the present level of 3 % to less than 1 % by October 1, 2000.
- With effect from May 1, 2000, all new buses to be purchased by BEST should be CNG operated until EURO II-compliant engines become available in these new vehicles. BEST may exercise an option either to have

(continued)

Box 2.2: (continued)

CNG-operated buses or EURO II or higher version diesel engine buses in such a manner that by April 1, 2005, at least 1,000 buses are operated on CNG.

- Engines of all the existing BEST buses, which are not even EURO I compliant, must be changed to EURO II-compliant engines by October 1, 2002.
- With effect from January 1, 2001, all taxis above the age of 15 years must be converted to CNG or any other clean fuel. Further with effect from January 1, 2002, all diesel taxis above the age of 8 years should be converted to clean fuel.
- With effect from January 1, 2001, all three-wheelers above the age of 10 years should be converted on CNG or any other clean fuel. Further with effect from January 1, 2002, all three-wheelers above the age of 8 years should run on clean fuel.
- The present permissible limit of 4.5 % CO emission in respect of two- and three-wheelers should be reduced to 3 % with effect from October 1, 2000, for Mumbai City to bring it at par with the CO emission level of four-wheelers.
- All heavy commercial vehicles as well as light good vehicles to be registered in the Mumbai Metropolitan Region from April 1, 2000 must be EURO II compliant.
- With effect from January 1, 2001, all two-wheelers registered in Mumbai Metropolitan Region and which are more than 15 years old shall be scrapped and their registration deemed to have been canceled.
- With effect from January 1, 2001, all three-wheelers registered in Mumbai Metropolitan Region and which are more than 10 years old shall be scrapped unless converted to clean fuel.
- With effect from January 1, 2001, all transport vehicles other than three-wheelers and BEST buses over the age of 15 years shall be scrapped unless converted to clean fuel.
- All two-stroke two- and three-wheeler in-use vehicles in Mumbai should be fitted with a catalytic converter by July 1, 2001.
- All petrol-driven vehicles registered in Mumbai prior to April 1, 1995, should be fitted with a catalytic converter by July 1, 2001.
- All catalytic converters supplied by the manufacturers for two-wheelers will carry a warranty of effective working of the catalytic converter over a distance of 30,000 km.

Source: CPCB (1999)

Apart from regulatory and judicial moves toward pollution control, there are a number of national initiatives. Two of such major national initiatives toward emission control are enforcement of supply of unleaded petrol (gasoline) and low sulfur diesel in the entire country, though implemented in phases. These initiatives implemented in April 1, 2000, and 1999 respectively have come out with visible positive impacts on environmental quality especially in Delhi.

Unleaded Petrol (*Gasoline*): The specification of lead in Indian petrol used to be 0.56 g/l max in 1994. Lead has been phased out from gasoline gradually, and by February 2000, only unleaded petrol is produced and sold in the entire country. In the developed countries, lead phasing out was spread over a period of 10–20 years. China and many other countries in the Asia-Pacific region and in Europe, South America, and Africa have not phased out lead as yet. However, India has achieved this within a span of 6 years.

Reduction of Sulfur Content in Diesel: This desulfurization program has resulted in the reduction in sulfur content by 75 % in 4 years. It has been brought down from 1.0 % to 0.25 % in the entire country in a period of 4 years (1996–2000). The Government had approved setting up of nine diesel hydrodesulfurization (DHDS) plants in nine refineries for reducing the diesel sulfur content at a total cost of Rs. 5,568.31 crore in June 1997. This has enabled supply of diesel with 0.25 % max. sulfur in the entire country from January 1, 2000. Further, in the four metro towns, sulfur content in diesel has been further reduced to 0.05 % (accounting to 95 % reduction). Detailed schedule of unleaded petrol and low sulfur program is presented in Box 2.3.

Subsequently, the Government of India has constituted a committee to look into transportation fuels in a comprehensive assessment, and the committee came up

Box 2.3: Gasoline Lead Phase-Out Program in India

Phase I	1.6.1994	Low leaded (0.15 g/l)	Cities of Delhi, Mumbai, Calcutta, and Chennai
Phase II	1.4.1995	Unleaded (0.013 g/l)	Cities of Delhi, Mumbai, Calcutta, and Chennai
Phase III	1.1.1997	Low leaded (0.15 g/l)	Entire country
Phase IV	1.9.1998	Ban on leaded fuel	NCT of Delhi
Phase V	31.12.1998	Unleaded (0.013 g/l)	All capitals of states/UT and other major cities
Phase VI	1.1.1999	Unleaded only	NCR
Phase VII	1.4.2000	Unleaded	Entire country

(continued)

Box 2.3: (continued)

Diesel sulfur phase-out program in India

Phase I	1.4.1996	Low sulfur (0.5 %)	Four metros and Taj Trapezium
Phase II	1.8.1997	Low sulfur (0.25 %)	Delhi and Taj Trapezium
Phase III	1.4.1998	Low sulfur (0.25 %)	Metro cities
Phase IV	1.4.1999	Low sulfur (0.25 %)	Entire country

Source: Basu (2001) and CPCB (1999)

with a detailed auto fuel policy (MoPNG 2002) prescribing a number of standards for Indian automobiles and also for different types of fuels.

With the above initiative, a reduction in vehicular emission load was observed and also improvement in ambient air quality.

With the increasing awareness and need to address the issue of urban transportation and the related issues, the Government of India has come up with National Urban Transportation Policy (NUTP) in 2006. NUTP outlines comprehensive guidelines for the State Governments in their preparation of transportation systems. NUTP is the first integrated approach toward urban transportation planning. The objective of this policy is to ensure safe, affordable, quick, comfortable, reliable, and sustainable access for the growing number of city residents to jobs, education, recreation, and such other needs within our cities. The details of NUTP are presented in Chap. 8.

References

Basu T (2001) Indian zero emission transportation programme – a driving force for change. In: Proceedings of international symposium and exposition on automotive electronics and alternate energy vehicles, 23–25 November 2001, IIT, Kanpur, pp 107–118

CPCB (1999) Parivesh, 6(1). Central Pollution Control Board, Ministry of Environment and Forests, Government of India, New Delhi

CPCB (2001) National ambient air quality status and trends in India – 2001, National Ambient Air Quality Monitoring NAAQMS/8,10,15,21

CPCB (2012) National ambient air quality status and trends in India – 2010, National Ambient Air Quality Monitoring NAAQMS/35/2011–2012

http.www.indiastat.com. Viewed on 1 Aug 2014

IGIDR (2001) "Techno-economic assessment of alternative options in urban transport sector for Delhi and Mumbai" under the project of analysis of technological alternatives for mitigation of GHG emissions from urban transport sector in selected Asian cities. IGIDR, Mumbai

Ministry of Statistics and Program Implementation (MoSPI) (2013) Statistical year book 2013. Ministry of Statistics and Program Implementation, Government of India
MoPNG (Ministry of Petroleum and Natural Gas) (2002) Report of the expert committee on auto fuel policy. Government of India
Yedla S (2004) Transportation trends, alternatives and policy issues. In: Radhakrishna, Parikh (eds) India development report 2004–2005. Oxford University Press

Chapter 3
Transport Service Provision: Environmental and Other Implications

3.1 Service Inadequacy and Congestion

Indian economic growth has been emphatic with growth rates consistent over and above 5 % for over a decade and reaching as high as 10.4 % (2010) (MoF 2014). Annual growth rate of vehicles (11 % in 1998–1991 and 9.9 % during 1991–2001) has been higher than the economic growth rate. As a result the number of vehicles has been increasing in number. This increase in ownership is shown in Figs. 3.1 and 3.2. The road transport service provision depends not only on the fleet of vehicles but also on infrastructure that is required by the other modes of transport. India has a very large fleet of vehicles with 141,866 million vehicles registered by the year 2011 needing a very large network of roads and large volumes of liquid and gaseous fuels. Maharashtra leads the list with a staggering 17 million vehicles followed by Tamil Nadu, Uttar Pradesh, Gujarat, and Andhra Pradesh. Among the metropolitan cities, Delhi leads the list with a very high vehicle population of about 7.5 million vehicles. However, the infrastructure, principally roads, did not increase at a comparable rate with the vehicular stock. Figure 3.3 presents the number of vehicles per 1,000 km road length. The road densities in urban and rural areas are 3,893.48 and 920 km/1,000 km^2, respectively. Road density per 1,000 population is 0.91 km in urban areas. As for percentage of total land area devoted to roads, Delhi has the highest at 21 % closely followed by Mumbai at 11 % and Kolkata at 5 %. This phenomenon has resulted in congestion on Indian roads and railways. The quality of roads has also been a problem with almost 50 % roads still uncovered/ surfaced. Figure 3.4 presents the quality of roads in Indian States.

Economic activity is emphatic in urban areas resulting in significant urbanization. The urban population in 2011 census reached 30 % of the total population in the country. This rapid economic growth and urbanization has resulted in densification of cities. The development area has more than doubled in cities. Vehicles per square kilometer land area during 1991–2005 has increased from 1,042 to 2,147 in Mumbai, 1,222 to 2,823 in Delhi, 543 to 1,684 in all metropolitan cities together,

S. Yedla, *Urban Transportation and the Environment*,
DOI 10.1007/978-81-322-2313-9_3

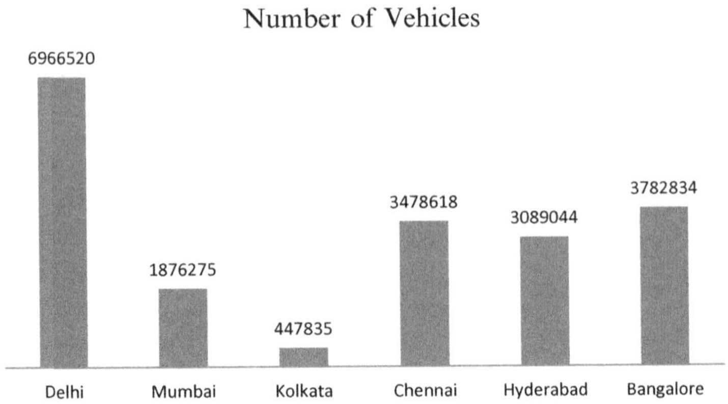

Fig. 3.1 Number of vehicles in four metropolitan cities in India (Source: Developed by the authors using data from Indiastat.com)

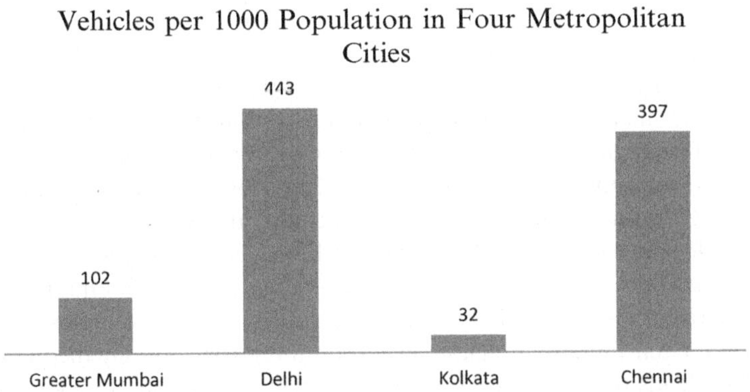

Fig. 3.2 Vehicle density (number of vehicles per 1,000 population) in four metropolitan cities in India (Source: Developed by the authors using data from Indiastat.com)

and 6 to 25 at all Indian levels. Inadequacy of infrastructure has resulted in crowded railways, congestion on roads, cramped city centers, inefficient energy consumption due to poor driving conditions, excess emissions, and overdependence on fossil fuels and the resulting greenhouse gas (GHG) emission.

Urbanization, as it happened in an unplanned way, has resulted in increasing sprawl in cities. As a result the Government has undertaken policies to decongest crowded city centers. Land-use regulation strictly limits the ratio of floor area to that of land area, and that restricts the heights of the building and the density of development in the city center. This leads to the development of suburban areas which have less restriction on the development. The suburban areas do not have a planned development either and suffer from not having required degree of connectivity with the mainland (Pucher et al. 2007). This process of decentralization has greatly affected urban transport. The expansion of cities has increased the length of trips of urban dwellers, leading to increase in the need to travel. Increased trip

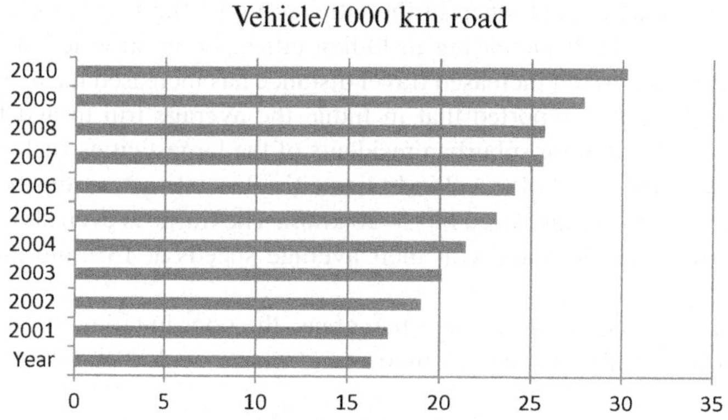

Fig. 3.3 Change in vehicle per 1,000 km road length over a period of 10 years (Source: Developed by the authors using data from Economic Survey of India (MoF 2014; Yedla 2014))

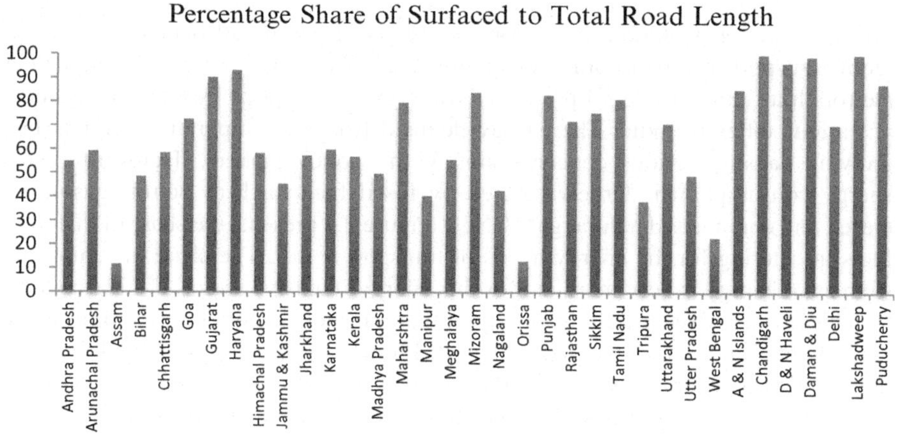

Fig. 3.4 Share of surfaced to total road length in different states of India

distances would further diminish the share of walking and cycling trips, encouraging a shift from non-motorized to motorized modes. As explained in previous chapters, a major share of work trips by the poorer sections of the society is by the non-motorized modes of transport (NMT). Among the smaller cities and towns, NMT serves the highest percentage of trips. This could be because of the low income and shorter travel distances in the tier II cities/towns. As city size increases, trip length increases as well, and that should be augmented by making public transport services more available. However, all the infrastructure development as the cities sprawl is focused on mobility of cars, and that is also not in proportion to the growth rate of automobiles. This results in:

1. Severe congestion in business districts and other city centers
2. Poor access for the suburban dwellers
3. Unaffordable travel service to disadvantaged sections of the society

Due to the urban sprawl wherein the cities are growing outwards, the average trip distances have been increasing in Indian cities. Congestion led slower travel speeds, and sprawl-driven increased travel distance has increased the average travel time significantly. It is reported that in India, the average trip to and from work requires up to 3 h a day for suburban residents of the largest cities such as Mumbai and Delhi. Average speed of travel in Indian cities has taken a serious beating with vehicles moving at average speed of 15–20 km/h. The traffic is even slower in cities such as Chennai and Kolkata with their average speeds at 13 km/h and 7 km/h, respectively.

As they are forced to live in suburban areas, the poor in cities end up spending 30–40 % of their daily earnings on their round trip to work (Peng 2005).

3.2 Energy Consumption

Transport sector is among the major consumers of energy all over the world. In India, transport sector consumes about 899 PJ of energy in the year 2005 among the metropolitan cities. Table 3.1 presents the energy consumption by transport systems of various cities in India. The energy demand from this important sector is fast growing causing serious concerns among the policy makers. Transport sector energy consumption is largely catered by fossil fuels such as petrol (gasoline), diesel, and compressed natural gas (CNG). Figure 3.5 presents the share of different fuels in catering for the energy needs of transport sector. The share of renewable sources of energy is very marginal for this energy-intensive sector. For the country to achieve energy security, it is important that transport sector energy footprints be made lighter and renewable energy based. Promotion of energy-efficient technology as well as energy-conserving modes such as mass rapid transit needs to be addressed on priority basis. Various transport management measures such as pollution under control certification, mandatory annual maintenance for vehicles completing their life span, etc.

Table 3.1 Energy consumption by transport systems of Indian cities

	Energy consumption from transport sector (PJ)		
Cities	1991	2001	2005
Delhi	26.84	41.82	48.01
Mumbai	13.89	15.04	16.97
Kolkata	5.1	8.63	10.19
Chennai	4.91	9.62	12.63
Hyderabad	5.47	10.84	11.51
Bangalore	7.91	15.96	18.06

Source: Compiled by the author by using data from MoSPI (2014)

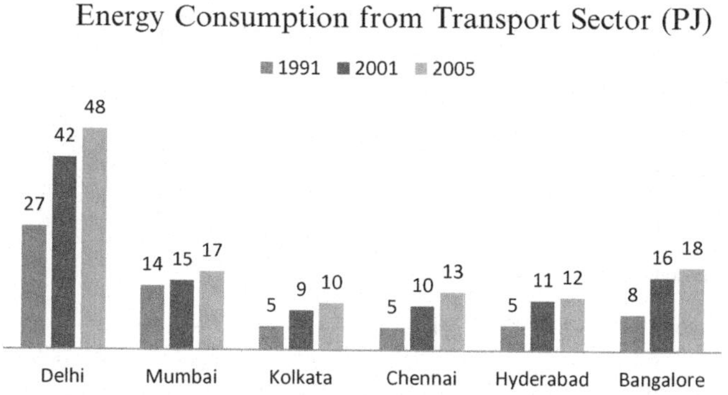

Fig. 3.5 Share of different fuels in transport sector (Source: Compiled by the author by using data from MoSPI 2014)

3.3 Safety/Accidents

Stressed transportation system results in traffic casualties and injuries, congestion, inadequate mobility, environmental pollution, and GHG emissions. Due to the overcrowding of roads, the total number of traffic-related accidents has gone up in the past two decades. Figure 3.6 presents the accident rate in India. Accidents are even more frequent in urban centers and also involve NMT users predominantly. Figure 3.7 explains the situation of transport causalities in the four major metropolitan cities in India.

Cramped vehicles result in safety issues, especially on trains and buses. During the peak hour of traffic as many passengers are forced to ride on the roofs or hang onto the sides of trains and buses, incidences of some passengers falling off overcrowded public transport vehicles are observed and pose an embarrassing side of the Indian transport system. Bicycles, motorcycles, and auto-rickshaws can also become dangerous when occupied by too many riders, which happens often in Indian metropolitan cities. The overcrowding of pedestrians, cyclists, and street vendors on the shoulders of roads creates additional safety problems, since they often spill over onto the roadway itself and are subjected to risks of road accidents.

The overall fatality rates per 100,000 motor vehicles is around 108 in India, more than five times as high as the fatality rate in the USA (18.5) in the year 2005. An academic estimate of traffic injuries in India is 1.2 million serious injuries and 5.6 million minor ones, though the official number is a lot lower (Pucher et al. 2007). The social dimension of traffic accidents is even more disturbing. It is the poor who use NMT that suffer the most in traffic fatalities. Such fatalities can have serious trickling down effects and devastating fanatical and social consequences to the immediate families.

The inadequate supply of roads, unsafe driving behavior, sharing of roads by motorized and non-motorized vehicles, overcrowding of vehicles, inadequate traffic signals and signs, and poor traffic management increase the danger of road

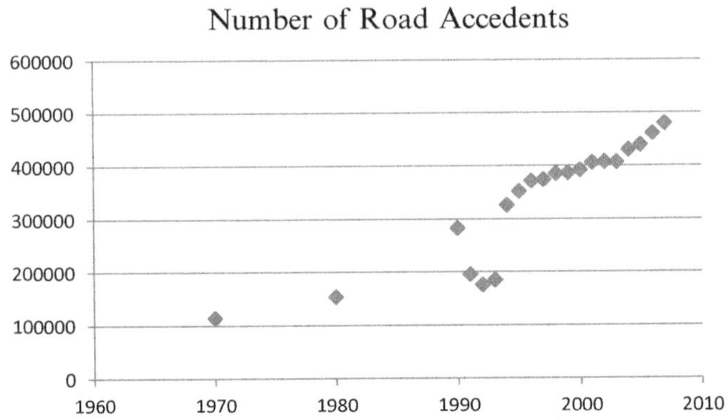

Fig. 3.6 Number of road accidents in India over a period of 1970–2007 (Source: Compiled by the author by using data from Indiastat.com and Ministry of Surface Transport and Highways)

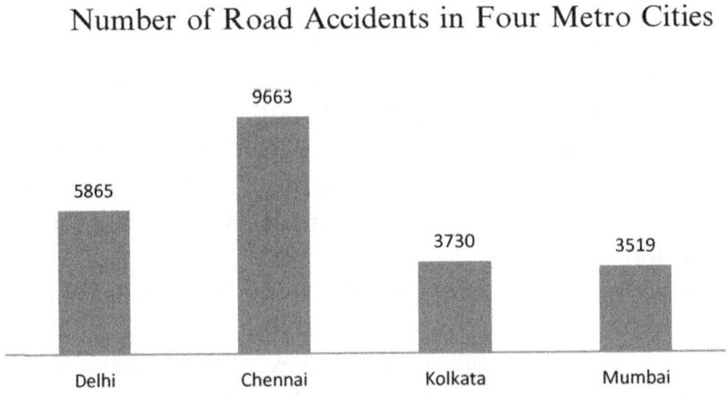

Fig. 3.7 Number of road accidents in India over a period of 1970–2007 (Source: Compiled by the author by using data from Indiastat.com and Ministry of Surface Transport and Highways)

catastrophes. Competing with the road users, street vendors occupy the footpath, forcing the walkers onto the road and making them vulnerable for accidents. Parking on roads is a common sight in India, and that takes away two effective lanes from the already narrow roads in urban centers. With such reduced road space, with increasing motorized vehicles, and being forced out of the footpaths by the venders, walkers are at extreme danger in Indian cities, and as reported by the World Bank, over half of all traffic fatalities involve pedestrians. With NMT users accounting for about 50 % transport fatalities, the urban poor are at highest risk. There is a general lack of modern traffic signals and signage to facilitate NMT users, and even where they exist, passengers often ignore them due to lack of enforcement by police. This not only reduces the speed of movement for the NMT users but also leads to increasing accidents in Indian cities.

3.4 Environmental and CO_2 Emissions and Noise Pollution

Transport sector contributes significantly to air pollution, noise pollution, and greenhouse gas (GHG) emission. With the increasing prominence of transport sector, air pollution has been on the rise and making many Indian cities among the most polluted in the world. Along with Beijing, Delhi is among the most polluted cities in the world.

Air pollution measurement programs over the last decade show a definite increase in average total suspended particles (TSP) and nitrogen oxide (NO_x) concentrations, while sulfur dioxide (SO_2) concentration has decreased (Pucher et al. 2007; CPCB 2010). TSP concentrations are much higher than WHO air quality guidelines of 90 g/m^3 at many sites in India. At certain times WHO air quality guideline for SO_2 is also exceeded. Mumbai, like many other cities, has a substantial particle pollution problem, with frequent and widespread exceeding of TSP and PM_{10} air quality guidelines (CPCB 2010). According to the measurements, the SO_2 pollution problem seems less pronounced although guidelines are sometimes exceeded. NO_x concentrations are presently within WHO guidelines. The higher levels of suspended particulates measured in Indian cities may be partly due to more dust in the air, from the dry surrounding countryside in northern India, from the many unpaved roads throughout the country, and from uncovered form of land use (Pucher et al. 2007).

India with its relaxed emission standards has evolved and now has one of the most stringent emission norms in the world. In spite of these norms, transport emissions contribute the major fraction of the overall air pollution. Table 3.2 presents the vehicular emissions vis-à-vis total emissions and explains the dominance of transport sectors' contribution to the overall emissions. Lead used to be a significant pollutant in Mumbai ambient air with annual average levels that ranged from 0.5 to 1.3 g/m^3. These exceeded the WHO guideline annual average (0.5–1 g/m^3, long term) at all locations of air quality monitoring in Mumbai. From 1980 to 1987, average lead concentration in the air nearly doubled. However, due to the directives of Supreme Court of India and subsequent implementation of "unleaded petrol," lead concentrations have reduced substantially.

Table 3.2 Vehicular emissions vis-à-vis total emissions in Greater Mumbai (ton/year)

	Vehicles	TSP	SO_2	NO_x
Gasoline	Cars	492	160	6,643
	Two-/three-wheelers	737	250	179
Diesel	Cars	765	395	1,783
	Buses	445	566	2,891
	Trucks	1,234	2,120	8,024
Transport sector		3,673	3,490	19,520
From all sectors		22,143	79,264	37,547

Source: WB (1997)

The summary of SO_2 measurements indicates that long-term average SO_2 concentrations are fairly low and less than WHO guidelines at all monitoring stations. It further improved with the directives for "low-sulfur diesel" program in Mumbai and all over the metro cities. Transport is responsible for more than 50 % of NO_x emissions.

While the new standards set by Indian government in the form of Bharat Stage I to IV in addition to the existing Euro standards, on road vehicles running on dirtier engines/technology pose a major challenge for the control of air pollution from India roads, there are still many older, more polluting motor vehicles on the road that must be converted or phased out. Inspection and maintenance is a potential tool to address this issue.

Air pollution from transport sector is further fostered by adulteration in automobile fuels. Auto-rickshaws, which are quasi-public and act as feeders to the mass transport and cater for the access and egress legs of a trip, illegally adulterate their gasoline fuel with up to 30 % kerosene and 10 % lubricating oil (Pucher et al. 2007). This is more of a socioeconomic problem in the society. However, with its impacts on the environment, such vehicle emits high concentrations of pollutants and cause severe air pollution. An additional problem in India is the large number of old motorcycles, scooters, auto-rickshaws, and tempos, many of which still rely on highly inefficient, poorly maintained, and very polluting two-stroke engines. The stop-and-go traffic flow caused by congestion also wastes energy and increases pollution.

Next only to the industrial source, transportation is a major source of noise pollution. With the rising motorization, the traffic noise has worsened in India, particularly so in Indian metropolitan cities. While the WHO standard for noise pollution is 45 dB, noise levels are much higher in Indian cities, with Mumbai and Delhi both registering more than 90 dB. That makes Mumbai the third noisiest city in the world, followed closely by Delhi (Garg et al. 2011). Ninety dB noise or louder can cause permanent shift of threshold and result to long-term hearing loss. Other smaller cities also are catching up with the metropolitan cities in traffic noise which is largely linked to the driving behavior in Indian cities and towns. Medium-sized Indian cities such as Nagpur, Varanasi, and Agra report noise levels ranging from 55 to 70 dB during peak hours.

3.5 ASIF Framework to Consolidate the Issues

Greenhouse gas emissions are the main cause of climate change (CC), and the transport sector contributes significantly to it. It not only contributed to the cause of climate change; it is also significantly vulnerable to the impacts of climate change such as a migration-driven complications and quality impairment due to the excessive rainfall and also raising groundwater levels.

Table 3.3 Greenhouse gas emissions from different sectors

Sector	CO_2 emission	CO_2 equivalence emission
Total national emissions (gigagram per year)	102,477.84	1,301,209.39
Energy and transformation industry	541,191.33	543,749.85
Industry	228,246.91	229,079.90
Transportation	95,976.83	98,104.12
Agriculture	–	355,600.19

Source: Compiled from India's National Communication to UNFCCC (MoEF 2012)

Carbon dioxide emissions from different sectors are given in Table 3.3. Urban transportation sector contributes to climate change by its CO_2 and methane emissions.

GHG emissions can be estimated by the following equation as developed by the author (Yedla and Garg 2014). The source of CO_2 emissions can be presented in a generic framework as shown below:

$$CO_2 = \underbrace{\sum_{i=1}^{m}\sum_{j=1}^{n}\sum_{k=1}^{o} EC_{ijk}}_{\text{Quantity of Consumption of Goods}} + \underbrace{\sum_{i=1}^{m}\sum_{j=1}^{n}\sum_{k=1}^{o} EP_{ijk}}_{\text{Quality of Production of Goods}}$$

where

CO_2 is the total CO_2 emissions
EC_{ijk} is the emission of CO_2 related to consumption from the sector i, using technology j and material k
EP_{ijk} is the emission of CO_2 related to production from sector i, using technology j and material k

Carbon emissions in specific to transportation can be calculated further by:

$$G = A{*}S_i{*}I_i{*}F_{ij}$$

where

G is the carbon emissions from transport sector
A is the total travel activity (PKM or TKM)
S_i is the model share of mode "i"
I_i is the energy intensity of mode "i"
F_{ij} is the fuel of type "j" in mode "i"

As explained above carbon emissions are linked to travel activity (A), type of travel modes used (S), energy intensity of the modes (I) used, and type of fuel (F)

used. In order to control emission from transport sector, it is important to address all the factors in integration. Benefits accrued by employing some of the measures can be easily offset by the other factors. For instance, the carbon emission control can be achieved by employing cleaner technology, and fuels can be offset by the increasing number of vehicles and the travel activity. The ASIF framework, developed by Lee Schipper et al. (2000), also helps in devising integrated strategies toward the planning of a sustainable transportation system.

References

Central Pollution Control Board (CPCB) (2010) National Ambient Air Quality Monitoring Series (NAAQMS/8,10,15,21)

Garg N, Sharma O, Maji S (2011) Noise impacts assessment of mass rapid transit system in Delhi. Indian J Pur Appl Phys 49:257–262

Ministry of Finance (2014) Economic survey of India 2013–2014. Ministry of Finance, Government of India

MoEF (Ministry of Environment and Forests) (2012) India – second communication to United Nations framework convention on climate change. Ministry of Environment and Forests. Government of India

MoSPI (2014) Energy Statistics 2014, Ministry of Statistics and Program Implementation, Government of India

Peng Z (2005) Urban transportation strategies in Chinese cities and their impact on urban poor, Paper presented at the Transportation Research Board 85th annual meeting, paper no. 05-2027

Pucher J, Peng Z, Mittal N, Zhu Y, Korattyswaroopam N (2007) Urban transport trends and policies in China and India: impacts of rapid economic growth. Transp Rev 27(4):379–410

Schipper L, Marie-Lilliu C, Gorham R (2000) Flexing the link between transport and greenhouse gas emission: a path for the World Bank. International Energy Agency, Paris

The World Bank (1997) Urban air quality management strategy in Asia, Greater Mumbai report. World Bank technical paper no. 381. The World Bank, Washington, DC

Yedla S (2014) Analytical paper on sustainable transportation and SDG – transportation in India: sustainability efforts and linkage to sustainable development goal, Report submitted to UNDP, New Delhi

Yedla S, Garg S (2014) Two decades of climate change negotiations – carbon budget allocation approach to re-shaping strategies. Int J East Asian Econ Integr 18(3):188–209

Chapter 4
Alternative Measures and Their Assessment

4.1 Review of Alternative Measures in Urban Transportation

4.1.1 Natural Gas Vehicles (NGVs)

The world NGV commercialization activities have taken place for varying reasons in different countries since their initial introduction in Italy in the mid-1930s. In India, it has started during 1990s. Italy is the first country in the world to use CNG as a fuel for transport, owing to its desire to become self-sufficient in fuels. Italians have been historic leaders in the NGV market, only recently outpaced by Argentina. Italians are product leaders worldwide, exporting vehicle conversion systems and compressor station equipment to the Middle East, South America, China, and India, to name a few countries. In other EU countries like Germany, though pipeline infrastructure for gas distribution exists, NGV markets are in the initial stages of development. In the USA, NGVs had their initiation in 1969 at the South California Gas Company, and now the Los Angeles Metropolitan Transport Authority has a fleet of about 2,600 buses, out of which 800 are NGV buses. All around the world, application of NG is predominant in buses though it has been used for cars as well. Delhi enjoys the distinction of having the highest share of CNG buses in the world with 51.5 % (2002) of its fleet running CNG, followed by Los Angeles with 30.5 % (parts of this chapter are derived from the material that has been published as a chapter in India Development Report 2002).

CNG vehicles are proved to be economically viable in many countries. In a study carried out by the author (IGIDR 2001), it was found that CNG technology with discount rate of 10 % outperforms the conventional technology. Table 4.1 presents the details. In another study carried out by TERI, the ratio of present net value of CNG cars over conventional cars was found to be less than unity, and the options

---

Some parts of the material presented in this chapter are derived from IGIDR (2001) and Yedla (2003).

S. Yedla, *Urban Transportation and the Environment*,
DOI 10.1007/978-81-322-2313-9_4

Table 4.1 Cost comparison of conventional vehicles vis-à-vis CNG vehicles in Delhi

Alternative	Life cycle operating cost[a] (Rs./passenger kilometer (PKM))[b]
Conventional three-wheelers (autos)	1.18
CNG three-wheelers (autos)	0.65
Conventional cars	2.45
CNG cars	1.92
Conventional buses	0.07
CNG buses	0.074

Source: Calculated by Author (IGIDR 2001)
[a]LCC includes the capital cost, fuel cost, operation and maintenance cost, etc.
[b]Discount rate, 10 %; year of estimate, 2001

of CNG are a financially attractive proposition even with a very high discount rate of 15 % with an assumed life of cars being 15 years or more.

However, a lack of an elaborate pipeline grid is a major barrier in many countries due to the NG economics becoming unfavorable, and promoting NGVs is tough, despite subsidies. In India, due to its vastness, natural gas vehicles can be commercially viable only in the cities where natural gas pipelines exist or would be laid in future. Currently CNG vehicles are prominent in major metropolitan cities like Delhi and Mumbai and slowly getting popular in other cities. Accounting for the external benefits from the clean environment might even favor the development of a new pipeline grid in other potential polluting cities like Hyderabad and Bangalore.

4.1.2 *LPG Technology*

Italy is the major user of LPG in the transport sector with 1.2 million vehicles running on LPG. Other countries using LPG as automobile fuels are Italy, the Netherlands, Poland, the USA, Canada, Mexico, Australia, New Zealand, Algeria, Turkey, Iran, South Korea, and Japan. In India, LPG is the main domestic fuel in urban areas. The indigenous availability of LPG is expected to fall much short of the household demand alone. For LPG to be competitive as an auto fuel, India would need Government support in the form of substantially lower taxation. Further, refilling and retrofitting are highly technology intensive that needs expertise in handling them. Safety poses another major concern for the use of LPG in transport.

4.1.3 *Electric Vehicles*

An electric vehicle (EV) is operated by an electric motor, which draws electricity from a battery. Storage batteries store a fixed amount of chemical energy.

The batteries can be recharged when the electrochemically active materials in these batteries have been used up. Because of availability and low price, lead acid batteries are more widely used than other batteries. In order to gain popularity, the electric vehicles should match with comparable petrol- or diesel-driven vehicles in terms of performance, reliability, durability, and cost. The useful life can be expressed in terms of number of cycles and in calendar periods. For the battery to be economical, it should have a minimum 1,000 full cycles, meaning that the battery can be fully charged and discharged at least 1,000 times before its capacity goes below 80 % of its rating. Go by the calendar period, the battery should last at least for 5 years.

India's efforts toward EVs have started with the Mashelkar committee (MNES 2000) set up by the Ministry of Non-conventional Energy Sources, Government of India, on high-energy density batteries for electric vehicles. As a result a number of electric vehicles are already plying on roads. BHEL had developed and commercialized a 16-seater EV Bus more than three decades ago. Three-wheeler electric vehicles are also commercially available in the country. Scooter India Limited has developed and demonstrated electric three-wheelers. Other major players for electric three-wheelers include Mahindra Eco Mobiles, which has come up with a three-wheeler EV called Bijli. Bajaj Auto and Eicher (Basu 2001) have also entered the three-wheeler EV market. Ashok Leyland has developed a hybrid EV. Reva, Bangalore, has developed battery-operated passenger cars. With further technical performance improvement, cost reduction, and increasing awareness, EVs are expected to find greater acceptance and market penetration.

The main targeted users of EVs would be public transport systems, urban services industries, service sectors, and Government utility vehicles. Exide Industries Ltd. at Kolkata and other manufacturers are producing lead acid traction batteries for EV applications. Lucas TVS Ltd. is also entering the EV area (Nair 2001).

In a financial viability study carried out by TERI (1997), it was found that the ratio of the present net value of battery-operated three-wheelers (autos) to that of conventional three-wheelers (autos) is less than unity for a discount rate of 6 % with life of 9 years. But the higher discount rates proved this technology uneconomical even at higher life span of vehicles. In a study carried out at IGIDR, it was found that BOV autos are more feasible than conventional autos in terms of life cycle operating cost. In spite of the higher initial investment, the life cycle operating cost of BOV was Rs. 0.63/PKM compared to that of conventional autos with LCC at Rs. 1.18/PKM. The discount rate assumed was 10 % and a life span of 9 years. Through the premier technological institutions in the country, the government has been supporting research on fuel cells. With improved technology, this attractive transport option has potential to play a key role in the years to come.

Table 4.2 Comparison of various auto technologies in India

Fuel type	Cost/km (Rs)	O & M (Rs)	Mileage (km/l)	Total cost/km (Rs)
Petrol	1.96	1.43	14	3.39
Diesel	0.58	0.80	20	1.38
CNG[a]				1.17
Battery-operated vehicle	1.64	0.32	16 kWh/90 kms	1.96
50 % concession on power tariff and concession on custom duty for battery	1.03	0.32	16 kWh/90 kms	1.35

Source: Extracted from Yedla (2004)
[a]LCC of CNG three-wheelers was taken from IGIDR (2001) where 10 % discount rate and 9 years of life span are assumed

4.1.4 *Hybrid Electric Vehicles*

Hybrid electric vehicles (HEVs) use the combination of an engine of a conventional vehicle with an electric motor powered by traction batteries and/or fuel cell. This combination helps in achieving both the energy and environmental goals. In hybrid electric vehicles propulsion, energy is available from more than one source of energy. The three configurations of HEV are series hybrid system, parallel hybrid system, and split hybrid system. A number of automobile manufacturers like Toyota, Honda, and DaimlerChrysler have started production of HEVs at different experimental levels, and some of them have also been produced on commercial scale.

Table 4.2 presents the cost comparison of various auto technologies in India. It clearly demonstrates that battery-operated vehicle (BOV) performs well under subsidies. CNG technology proved to be doing better than BOV.

4.1.5 *Fuel Cells*

In an electrochemical process, fuel cells produce electricity, employing reaction between hydrogen and oxygen gases. Fuel cells are efficient, environmentally benign, compact, modular, and reliable for power generation. The different types of fuel cells currently under development are the proton exchange membrane fuel cells (PEMFCs), phosphoric acid fuel cells (PAFCs), molten carbonate fuel cells (MCFCs), solid oxide fuel cells (SOFCs), direct methanol fuel cells (DMFCs), and alkaline fuel cells (AFCs).

Fuel cells operate on hydrogen gas and oxygen from air. Hydrogen gas can be obtained from a variety of fuels. Fuel cell power plants can be operated with overall system efficiencies of about 45–60 % or even higher. They offer high efficiency and low emission. Because of high fuel utilization efficiencies, fuel cells help in carbon dioxide emission reductions. Fuel cells can be used for stationary/portable power generation and automotive applications. High conversion efficiency, extremely low or no emissions, noiseless operation, high current density, and compactness are some of the advantages which make fuel cells an ideal power option for automobile application. The use of fuel cells has been demonstrated in transport applications. For regular operation of fuel cell electric vehicles, refueling infrastructure needs to be created. A number of firms such as Ballard Power Systems, Plug Power, Fuel Cell Energy, International Fuel Cells Corporation, Energy Partners, H-Power, Honeywell, Northwest Power Systems, Toyota, Siemens, and others have been targeting fuel cell markets for transportation.

In India, several agencies, and principally MNES, have been supporting projects on different types of fuel cell technology. Small PAFC stacks have been developed and tested by BHEL.

4.1.6 Hydrogen Energy

In the times of climate change, hydrogen is receiving worldwide attention as a clean fuel and efficient energy storage medium for automobiles. Hydrogen can replace or supplement oil used in road transportation. Hydrogen production technologies can be classified as fossil fuel based and renewable resource based. Hydrogen gas is being produced on a commercial scale by steam reformation and thermal cracking of natural gas or partial oxidation of naphtha, conventional electrolysis, and coal gasification. Hydrogen can also be produced from steam reforming/thermal cracking of natural gas, partial oxidation of naphtha, and conventional electrolysis of water.

The cost of production of pure hydrogen from the electrolysis of water using photovoltaic device needs to be reduced substantially, before it becomes economically viable. Storage, transportation, and delivery are of great importance as it involves considerable hazard. In India, as part of the R&D project, a photobioreactor for producing hydrogen from distillery waste treatment was installed at Nellikuppam in Tamil Nadu. R&D projects were also taken up for the production of hydrogen using solar energy and water through photoelectrochemical/photocatalytic methods. Technologies are being incubated and it would be a while before they hit the market.

4.1.7 Biofuels

Biofuel is an efficient, environment-friendly, natural energy alternative to petroleum fuels. In view of the potential of being produced from several agricultural sources and because of its low emission characteristics, biofuels in recent years are receiving attention as a substitute to fossil fuels. Ethanol and biodiesel are the two biofuels which are being looked upon as the potential fuels for surface transportation.

Ethanol, which is being used mainly as a raw material for chemical industries, in medicines, and for potable purposes, is being increasingly looked upon as a potential fuel for powering automobiles. This can be obtained through the processes of fermentation and distillation from molasses, a by-product of sugar industry. Other possible raw materials are cassava, corn, rice straw, and potato. The Government of India has already taken initiative to blend 5 % ethanol in petrol on a commercial scale in a phased manner. Further R&D work on the use of ethanol in diesel fuel has also been taken up.

The term "biodiesel" refers to the neat ethyl esters of vegetable oils. Presently absolutely pure or neat methyl esters of rapeseed, soya bean, sunflower, talon, and other fats and oils are used as diesel fuel without any substantial modification to the existing design of the engine. Biofuel is an ideal synergistic partner for oxidation catalytic converter and reduces CO_2 emissions by 78 % when compared to conventional diesel fuel. According to the Economic Survey, Government of India, out of the cultivable land area, about 175 million hectares is classified as waste and degraded land. Therefore, there is a good potential to tap this resource for the production of biodiesel.

4.1.8 Inspection and Maintenance (I&M) and Certification System

Vehicle inspection is a tool employed to ensure that vehicles meet applicable exhaust emission standards under normal operating conditions. Two different types of inspections are commonly performed:

- Periodic inspection in which the vehicle is inspected at an approved testing station at regular fixed time interval (i.e., every 3/6 months)
- Spot inspection performed on a random basis by pulling vehicles off the road for on-site inspection ("roadside inspection")

If testing determines the vehicle is exceeding the emission requirements, the vehicle owner is required to perform the necessary corrective actions within a prescribed time period. Fines or withholding registration tags are normally employed to ensure that the vehicle owner carries out the necessary corrective action. Any I&M program requires a considerable amount of investment not only to set up the facilities but also for tests.

4.1.9 Retrofitting of Emission Control Devices

Even after enforcing a thorough and reliable I&M program for in-use vehicles, the emission level of the vehicles cannot go below the levels for which the vehicle has been designed. The successful implementation of a retrofit program for reducing emissions would depend on necessary engine modification, in conjunction with the retrofitting of emission control devices, which have adequate durability. There should be a sustainable and widespread availability of the required quality of fuel. Active participation of various regional transport authorities, vehicle manufacturers, and fleet operators is essential to make this process effective.

4.1.10 Traffic Management and Construction of Bypasses

Traffic management has great potential to control traffic growth and resulting environmental emissions. Following are a few management measures that can be used to improve transportation efficiency and also its environmental performance.

- Installation of traffic control devices at intersections
- Optimization of signal cycle timings and synchronization of traffic signals
- Different cycle timings for different periods
- Provision of adequate road geometric and flyovers and free left turns
- Increased mass transport system
- Diversion of vehicles to less crowded routes and closing streets at peak times selectively
- Segregating fast-moving and slow-moving traffic by means of designated lanes
- Physical restraint of traffic by closing roads, one-way systems, etc.
- Restrictions on personalized low-occupancy vehicle use on specified days/times/roads
- Car pooling
- Prohibiting single-passenger personalized vehicles at peak times on congested roads
- Congestion charges on crowded roads
- Parking charges that vary with vehicle type and time of day and location

4.1.11 Construction of Metro Rail, Monorail, and BRTS

Urban transportation is at crossroads with problems such as congestion, long travel distances, slow movement of traffic, and air pollution and greenhouse gas (GHG) emissions coexisting. After exhausting retrofitting measures such as flyover bridges and catalytic converters, the Government of India has adopted measures to augment mass transport. While initiatives on augmenting buses are going at a slow phase, development of capital intensive modes such as metro rail, monorail, and BRTS has been gaining importance as well as faster implementation. These mass modes in combination with bus network can potentially provide a solution for the travel needs.

Table 4.3 Details of monorail projects being planned for Indian cities

Project	City	Cost (crores)	Status
Bangalore monorail project	Bangalore	7,700	Planning
Thane-Bhiwandi-Kalyan monorail project	Thane	3,750	Planning
Thiruvananthapuram monorail project	Thiruvananthapuram	3,590	Planning
Jacob Circle-Wadala-Chembur monorail project	Mumbai	3,000	Partially completed
Delhi monorail project	Delhi	2,235	Planning
Kozhikode monorail project	Kozhikode	1,991	Planning
BKC monorail project	Mumbai	1,125	Planning
Chandigarh monorail project	Chandigarh	NA	Planning
Chennai elevated monorail MRTS project	Chennai	NA	Planning
Port Blair monorail project	Port Blair	NA	Planning
Dehradun monorail project	Dehradun	NA	Planning
Pune monorail project	Pune	NA	Planning

Metro rail has already been operating in cities such as Kolkata, Delhi, Bangalore, Mumbai, and Jaipur and also in construction stages in Chennai, Kochi. More cities such as Pune and Visakhapatnam are planning for their own flashy transport system.

Monorail is also being planned in many Indian cities. Mumbai is the first city to have a monorail, and Delhi, Bangalore, Thiruvananthapuram, Kozhikode, Chandigarh, Chennai, Port Blair, Dehradun, and Pune are planning to have the monorail system. Table 4.3 presents the details of the proposed monorail project in India.

Bus rapid transit system (BRTS) is another public transportation mode promoted by the Government of India in different cities. The Central Government has sanctioned bus rapid transit system (BRTS) projects to 11 cities, and details include seven projects/phases of BRTS in Pune (Maharashtra); three each in Ahmedabad (Gujarat), Indore (Madhya Pradesh), and Jaipur (Rajasthan); two in Bhopal (Madhya Pradesh); and one each in Vijayawada and Visakhapatnam (Andhra Pradesh), Rajkot and Surat (Gujarat), Amritsar (Punjab), and Kolkata (West Bengal). More than 500 km is to be covered under these BRTS projects sanctioned by the Ministry of Urban Development.

4.2 Methodological Approach for the Assessment of Alternatives

While planning for the urban transportation, it is important to assess all the available alternatives for their potential in saving energy and control over emissions along with the other attributes. Following are the set of social and economic conditions that any alternative chosen for the betterment of urban transport should satisfy (Turner et al. 1994):

1. Environmental effectiveness/emission reduction potential
2. Energy-saving potential
3. Economic viability by means of life cycle operating cost (LCC)
4. Pollution abatement cost and marginal abatement cost
5. Equity
6. Low administrative costs
7. Institutional compatibility and concordance

This section presents the methodology that can be used to assess the abovementioned conditions with particular emphasis to emission reduction potential (ERP), energy-saving potential (ESP), economic viability by means of LCC, pollution abatement cost (PAC), and marginal abatement costs (MAC).

4.2.1 Energy-Saving Potential

The energy-saving potential of an alternative option is presented as its relative energy requirement compared to the energy consumption of the existing mode of transport. Energy-saving potential can be calculated in the following steps:

Step I: $\mathrm{PKM}_{jt} = V_{jt}{}^{*}U_{j}{}^{*}365$

where

V_{jt}, number of vehicles of type "j" in the year "t"
U_j, average daily utilization of vehicle type "j"
PKM_{jt}, annual passenger kilometer covered by vehicle type "j" in the year "t"

Energy requirement of each option can be determined by considering the total PKM catered by respective mode of transport of a particular option and the respective energy intensity factor. The following equation could be used to determine the energy demand:

Step II: $F_{jt} = \mathrm{PKM}_{jt} \times E_j$

where

F_{jt}, energy requirement of the vehicle type "j" (alternative/existing) in the year "t"
E_j, energy intensity of the vehicle type "J" expressed in fuel consumption (GJ) per PKM

Then energy-saving potential of the alternative option "o" can be determined as follows:

Step III: $(\mathrm{ESP})_o = 1 - \frac{F_{jt_{\mathrm{alt}}}}{F_{jt_{\mathrm{old}}}}$

where

ESP_o, energy-saving potential of alternative option "o"
$(F_{jt})_{old}$, energy consumption by the conventional vehicle type "j" in year "t"
$(F_{jt})_{alt}$, energy consumption by the alternative vehicle type "j" in year "t"

Unity of this index represents a better potential of the alternative option.

4.2.2 *Emission Reduction Potential*

One of the major objectives of the alternative measures in transport sector is the reduction in air pollutants such as CO_2, CO, SO_x, NO_x, HC, TSP, and Pb. Emission reduction potential can be calculated in the following steps:

Step I: $P_{ijt} = F_{jt} \times EF_{ijt}$

where
P_{ijt}, emission of pollutant "i" in tons by vehicle type "j" in the year "t"
F_{jt}, energy requirement of the vehicle type "j" (alternative/existing) in year "t"
EF_{ijt}, emission factor for pollutant "i" of vehicle type "j" expressed in kg per kg of fuel used in year "t"

Step II: $(\mathrm{ERP})_{oi} = 1 - \dfrac{P_{ijt_{alt}}}{P_{ijt_{old}}}$

where

$(P_{ijt})_{old}$, emission of pollutant "i" in the conventional case of vehicle type "i" in year "t"
$(P_{ijt})_{alt}$, emission of pollutant "i" in the alternative case of vehicle type "j" in year "t"
ERP_{oi}, emission reduction potential

Given the equal importance (weights) for all pollutants in a particular alternative option, the overall compounded emission reduction potential of option "o" can be expressed as follows:

Step III: $$\mathrm{ERP}_o = \frac{\sum_{i=1}^{n}\left[1 - \dfrac{(P_{ijt})_{alt}}{(P_{ijt})_{old}}\right]}{n}$$

where
$P_{i-\text{alternative}}$, emission of pollutant type "i" in the alternative technology in the year "t"
$P_{i-\text{old}}$, emission of pollutant type "i" in the existing technology in the year "t"

$$Step\ IV: \mathrm{ERP}_o = \sum_{i--=1}^{n} \left\{ \left[1 - \frac{(P_{ijt})_{\mathrm{alt}}}{(P_{ijt})_{\mathrm{old}}} \right] * W_i \right\}$$

where

W_i, weight for pollutant i

The higher the values of ERP, the better the potential of the alternative to save energy. It is possible to get values more than unity in the case when the new technology results in an increase in pollution (it may be an economically viable option). This kind of quantitative approach would facilitate comparison among different potential options. Unit emission of all local pollutants, both for existing and alternative technologies, can be considered in making a judgment over the alternative options and their selection.

4.2.2.1 Data Input and Assumptions

Data requirement includes the travel demand met by different models of transport, energy demands of various transport modes, and pollutant emissions.

While calculating the pollution per the unit output, only the mode of transport involved should be considered. For instance, for the option of "use of CNG for buses," the pollution generation by the existing and alternative options per unit output by buses needs to be calculated for different pollutants.

4.2.3 *Economic Viability*

Economic viability is an important requirement for any development activity to be successful. Literature demonstrates the fact that any development option, irrespective of its technical potential, tends to fail if it is not a cost-effective option. Hence, for the selection of various potential pollution mitigation options from transport sectors, economic analysis needs to be carried out for all the options that are selected based on their potential for emission reduction.

The costs need to be considered throughout the lifetime of each option to determine the present value and also the net present value. Levelized cost should be determined for each option by considering the present value and certain discount rate (excluding) and life span of the vehicle. Taking annual passenger kilometer covered by the respective mode in to account, life cycle operating cost (LCC) can be determined for each option. Methodology for the determination of LCC is presented below.

4.2.3.1 Levelized Costs (LC) of the Selected Options

Levelized cost is the concept used in economics to determine the cost of the system distributed throughout its life span. The net present value can be determined by taking all the cost, viz., capital, operation, and maintenance, fuel costs, taxes, etc. Subsidy on fuel should be factored out. Net present value is used to determine the levelized cost of each option under consideration. Following is the brief methodological note on determination of LC.

Levelized cost can be determined using the following equation:

$$\text{Levelised cost (LC)} = \text{PV}^* \left[\frac{i}{1-(1+i)^{-n}} \right]$$

where

PV, present value
i, discount rate
n, life span

4.2.3.2 Life Cycle Costs (LCC) of the Alternative Options

Life cycle cost is the operating cost involving per unit output of the alternative option expressed in terms of cost/unit output (km or pkm). The following equation can be used to calculate the life cycle operating cost:

$$\text{Life cycle operating cost} \quad \text{LCC} = \frac{\text{LC}}{\text{PKM}}$$

where

LC, levelized cost of the option
PKM, annual passenger-km traveled = annual utilization × load factor of the mode

4.2.3.3 Pollution Abatement Cost (PAC)

Potential of the alternative options in controlling emissions in an economically viable means can be determined by computing pollution abatement cost. Per unit of pollution abatement cost (PAC) is represented by the additional cost, which would be incurred in introducing an environment-friendly technology, i.e., which would result in reduction of emission from the transport sector. This is derived in terms of cost/ton and is given by the following equation:

$$\text{PAC} = \frac{\text{LCC}_1 - \text{LCC}_0}{E_0 - E_1}$$

Table 4.4 Characteristics of PAC values

PAC	Conditions
Positive (+ve)	Technology is costly and less polluting
Negative (−ve)	Technology is cheaper and less polluting
Positive (+ve)	Technology is costly and more polluting
Negative (−ve)	Technology is cheaper but more polluting

where

PAC, per unit of pollution abatement cost
LCC_0, life cycle cost of the existing technology
LCC_1, life cycle cost of the alternative technology
E_0, emission per unit output of the existing technology
E_1, emission per unit output of the alternative technology

PAC tends to get negative values if the alternative technology is cheaper than the existing one and also less polluting. This attains a positive value if the alternative technology is costlier than the existing one and also less polluting. However, it is possible to get PAC negative if the alternative technology is costly as well as more polluting. This is possible with some alternatives that result in certain nontarget increasing pollutants. In such a case, the PAC attains a positive value when the technology is cheaper but is more polluting. Table 4.4 demonstrates various possible PAC characters.

4.3 Empirical Presentation

The methodology developed is applied to a set of alternatives selected from Delhi transport system in order to demonstrate the process of assessing transportation alternative.[1]

Delhi transport is dominated by two-wheelers and cars. Two-wheelers occupy 67 % of the Delhi passenger fleet followed by cars (29 %). Buses, taxis, and three-wheelers occupy the remaining fraction of vehicular fleet. Delhi traffic is driven predominantly by two-wheelers and cars. Table 4.5 presents the number of vehicles and PKM catered by each mode of transport in Delhi. Domination of two-wheelers and cars can be clearly observed from the table with respect to the number of vehicle. Buses dominate the PKM catered.

Taking this fact into consideration, three alternative options are considered for the analysis. The methodology described in the previous section is applied to assess these alternatives considered for Delhi transport system. Alternative options considered are:

[1] The results presented in this section have been published in parts in Yedla (2001b) and Yedla (2001a).

Table 4.5 Details of number of vehicles and PKM covered in Delhi

Travel mode	Number of vehicles	Travel demand met (billion PKM)
Two-wheelers	153,0944	11.32
Three-wheelers	5,9831	4.61
Cars	66,4564	17.55
Taxis	12,802	0.62
Buses	15,532	39.02

Source: Compiled by the author using data from Indiastat.com

1. Conversion of all 2-stroke two-wheelers into 4-stroke two-wheelers
2. Conversion of all cars running on conventional fuels with CNG
3. Conversion of all buses running on conventional fuels to CNG

4.3.1 Data Requirements and Assumptions

Assessment of alternatives in urban transport would require the following data:

- Energy efficiency by different vehicle type/technology
- Emission factors of various pollutants for different vehicle types/technologies
- Utilization of the vehicles
- Load factors for different vehicles
- Travel demand in PKM
- Capital costs of various vehicles both of existing and alternative technologies
- Fuel costs of various vehicles both of existing and alternative technologies
- Operation and maintenance costs of different vehicles of different technologies
- Various other costs involved in vehicle operation
- Discount rate
- Life of vehicles of different technologies

Energy demand and the respective emissions of different pollutants for all the three options are calculated using the formulation described in the previous section. All three options showed significant influence on different pollutants and their levels in overall pollution levels in Delhi. Table 4.6 presents the pollution of each pollutant per PKM traveled in the respective mode of transport for each option. Emission reduction potentials of all alternative options for different pollutants are calculated using the formulas given in the methodology. Table 4.7 presents ERP of different options for different pollutants.

Alternate option I (conversion of two-wheelers) showed significant influence on carbon dioxide emissions with 31 % reduction in emissions. It further showed its potential in reducing TSP (84 %), hydrocarbons (86 %), and SO_x (31 %). However it has resulted in considerable increase in NO_x emissions. Potential to reduce emission of lead is less as compared to other options. Alternative option II (CNG cars) which is friendly for local environment contributes to global environmental

Table 4.6 Emission reduction of each mode of transport option per unit output

Option	Fuel type	Emission (g)/PKM						
		CO_2	CO	SO_x	NO_x	HC	TSP	Pb
Two-wheelers	2-stroke	37.70	4.53	0.0257	0.0545	2.8251	0.2726	0.002
	4-stroke	25.83	4.53	0.0177	0.2128	0.3939	00437	0.001
Cars	Gasoline	73.19	3.95	0.1306	0.5495	0.3833	0.1164	0.002
	CNG	78.01	0.0042	0	0.0669	0	0	0
Buses	Diesel	22.89	0.3055	6E-05	0.5054	5E-5	4.8E-5	0
	CNG	35.84	0.0019	0	0.0307	0	0	0

Table 4.7 Emission reduction potential of different alternatives

	Emission reduction potential						
Option	CO_2	CO	SO_x	NO_x	HC	TSP	Pb
4-stroke two-wheelers	0.3148	−0.0008	0.315	−2.904	0.861	0.839	0.313
CNG cars	−0.066	0.998	1.000	0.878	1.000	1.000	1.000
CNG buses	−0.565	0.994	1.000	0.939	1.000	1.000	0.000

Table 4.8 Priorities of different alternatives based on ERP

Option	Weighted ERP
4-stroke two-wheelers	0.2277
CNG cars	0.9876
CNG buses	0.8933

problem. It increases CO_2 emissions by 7 % while reducing CO emissions by 99.89 % and NO_x by 87 %. Potential of this option in reducing TSP and HC is as high as 100 %. Alternative option III followed a similar trend as option II except that the CO_2 contribution increased significantly. This option of converting diesel buses to CNG buses resulted in increase of CO_2 emissions by 57 %.

As all these options show different potential for different pollutants and their reduction, it would be highly difficult to choose or prioritize these options unless they are handled by multi-criteria optimization. Table 4.7 presents emission reduction potential (ERP) calculated based on the methodology developed in the previous sections. A similar scenario is reflected by ERP in a much simpler way but still leaving the complication of integrating these multiple pollutant ERPs. To arrive at a single index which presents the pollution reduction potential, weights are attributed to various pollutants given their importance to the particular case under study. TSP, HC and SO_x concentrations typically dominate Delhi air pollution. Therefore it is appropriate to assign weights for the mitigation potential for different pollutants for each option. Different procedures can be followed for weight derivation for different pollutants. In the present case, environmental experts' opinion is considered for the sake of simplicity. CO_2 is not considered in the process of weighted ERP. Overall emission reduction potential of each alternative option has been calculated and is presented in Table 4.8.

Common emission reduction potential (CERP) gives an index of emission reduction potential for any particular option. The priorities that are calculated based on final ERP provide a clear ranking of options. From the table it is clear that option I performed low on controlling local pollutants as compared to other two options.

4.3.2 *Cost-Effectiveness of the Alternatives*

The three alternative options which showed significant ERP are tried for their economic feasibility. Ten percent discount rate is considered for all NPV calculations. Life span of bus is taken as 20 years.

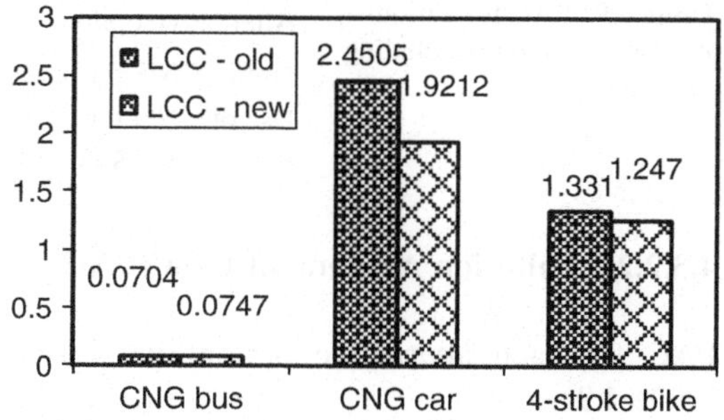

Fig. 4.1 Life cycle operating cost of various alternative options

Conversion of buses running on conventional fuels with CNG showed a levelized cost (LC) of 169,768 (2,829 USD at Rs.60/USD) over the LC of the existing buses, Rs. 159,937 (2,665 USD). The initial cost is so high in the case of CNG buses and it added up to the LC of this option in spite of a large difference in fuel costs.

The LC for the CNG cars is Rs. 36,387.5 (606 USD) compared to the LC of Rs. 46,397 (773 USD) for the existing cars and taxis. Unlike the earlier option, this option showed a better LC for the alternative option. This may be due to the fact that the retrofitting charges for the CNG kit is not very high and also there is a major saving in the fuel cost. But the conversion charges are very high (as high as 40 % of the capital) in the earlier option. This must be the reason for the high LC of the alternative option over the existing one.

Four-stroke two-wheelers showed a better LC over the existing two-stroke two-wheeler. The LC of the alternative option is Rs. 8,332 (138 USD) compared to that of the base case which is Rs. 8,893 (148 USD). LC is low to a magnitude of 8 %.

4.3.2.1 Life Cycle Costs (LCC) of the Selected Options

In the case of Delhi, CNG bus showed higher LCC than that of the existing diesel bus. The LCC of the CNG bus is 0.07 Rs/pkm compared to that of the existing bus which is 0.70. LCC of the existing car system is 2.45 compared to that of CNG car which is low at 1.92. This may be due to the fact that retrofitting cost of CNG kits is not very high and the fuel savings are considerable unlike in the case of buses. However, it is interesting to notice that overall life cycle operating cost is much higher for cars in comparison to buses. Replacing 2-stroke two-wheelers by 4-stroke two-wheelers showed a promising trend in LCC with a value of 1.25 over 1.33 for 2-stroke bikes. LCC of all three alternative options are shown in Fig. 4.1.

As it can be observed from the above figure, LCC of car, which is a personalized mode of transport, is much higher than the other public mode of transportation. The variation of LCC was found very insignificant for the case of buses, which is a cheaper mode of transport in comparison to two-wheelers and cars.

Table 4.9 PAC for various alternative options (including all pollutants)

Alternative option	PAC (Rs/g)
Use of CNG for buses	0.00045
Use of CNG for Cars	−0.1044
Replacing 2-S bikes by 4-S	−0.0335

4.3.2.2 Pollution Abatement Cost (PAC)

PAC for each alternative option that is calculated for Delhi is presented in Table 4.9.

Cars are the most expensive mode of transport in Delhi, followed by two-wheelers and buses. It is also observed that the difference in LCC for the existing technology and alternative technology is very narrow in the case of buses. The PAC analysis proved that it is expensive to reduce pollution using the CNG bus. CNG car is a much cheaper mode to reduce pollution than that of a shift to 4-stroke bikes. From the above table, it is apparent that CNG cars are doing well followed by 4-stroke bikes in controlling pollution in a cost-effective way.

References

Basu T (2001) Indian zero emission transportation programme – a driving force for change. In: Proceedings of international symposium and exposition on automotive electronics and alternate energy vehicles, 23–25 November 2001, IIT, Kanpur, pp 107–118

IGIDR (2001) "Techno-economic assessment of alternative options in urban transport sector for Delhi and Mumbai" under the project of analysis of technological alternatives for mitigation of GHG emissions from urban transport sector in selected Asian cities. IGIDR, Mumbai

Ministry of Non-conventional Energy Sources (MNES) (2000) Report of Mashelkar's Committee on development of high energy batteries for electric vehicles. Ministry of Non-conventional Energy Sources, Government of India

Nair KRA (2001) Side mounted starter-generator for small size passenger cars. In: Proceedings of international symposium and exposition on automotive electronics and alternate energy vehicles, 23–25 November 2001, IIT, Kanpur, pp 170–176

Tata Energy Research Institute (TERI) (1997) Environmental aspects of energy use in large Indian Metropolis, vol I, New Delhi

Turner RK, Pearce D, Bateman I (1994) Environmental economics – an elementary introduction. Harvester Wheatsheaf, New York

Yedla S (2001a) Application of analytic hierarchy process for the selection of alternative options for urban transportation in Delhi – analysis of group aggregation methodology adopted in AHP, Post-doctoral Fellowship Report, AIT, Bangkok, Thailand

Yedla S (2001b) Index development for emission reduction potential of various alternative transportation options: methodology and application. In: Proceedings of fifth international conference of the International Society for Ecological Economics (ISEE) Russian Chapter (Russian Society for Ecological Economics - http://RSEE.narod.ru/)

Yedla S (2004) Transportation trends, alternatives and policy issues, In: Radhakrishna and Parikh (eds) India development report 2004–2005. Oxford University Press

Chapter 5
Non-motorized Modes of Transport

5.1 Present Scenario of Non-motorized Transport

The definition of NMT includes any form of transportation that provides personal or goods mobility by methods other than the combustion motor (Guitink et al. 1994). Walking is the most common form of non-motorized transportation (NMT), and the other forms of NMT include bicycles, tricycles, human porterage, handcarts, animal-drawn carts, and other human-powered vehicles.

In the early years of the last century, bicycles used to have a major share of daily trips. With the objective of speeding up the movements, motor vehicles gained popularity over a period of time. With the economic progression, the motor vehicles have slowly taken over completely sidelining the non-motorized modes of transport. Overemphasis on motorized transport in the funds allocation for infrastructure development had further jeopardized the prospects of NMT. However, this trend is predominant in urban centers only, leaving bicycles still a major source of mobility in the rural areas. This particular trend is predominant in developing countries, whereas in the developed countries, the motorization had penetrated into rural areas as well.

In many Asian countries, non-motorized bicycles (two-wheelers) and three-wheelers are common sights, employing special adaptations to the vehicle for goods transport, hawking, or taking passengers. Walking accounts for two-thirds of total trips in large African cities such as Kinshasa and Dar es Salaam. In Karachi, walking and cycling account for 60 % of total trips and for 40 % of work-related trips. In Chennai, India, one-third of the vehicles entering the central business district are bicycles. In most of the Chinese cities, 50–90 % of vehicular passenger movements are by bicycle, with most of the remainder by bus.

In Mumbai 40–60 % of daily trips are made by NMT (Cropper and Bhattacharya 2007; Rastogi 2002; Tiwari 2003). However, use of bicycle is insignificant in

Parts of the material presented in this chapter are extracted from Yedla (2007).

S. Yedla, *Urban Transportation and the Environment*,
DOI 10.1007/978-81-322-2313-9_5

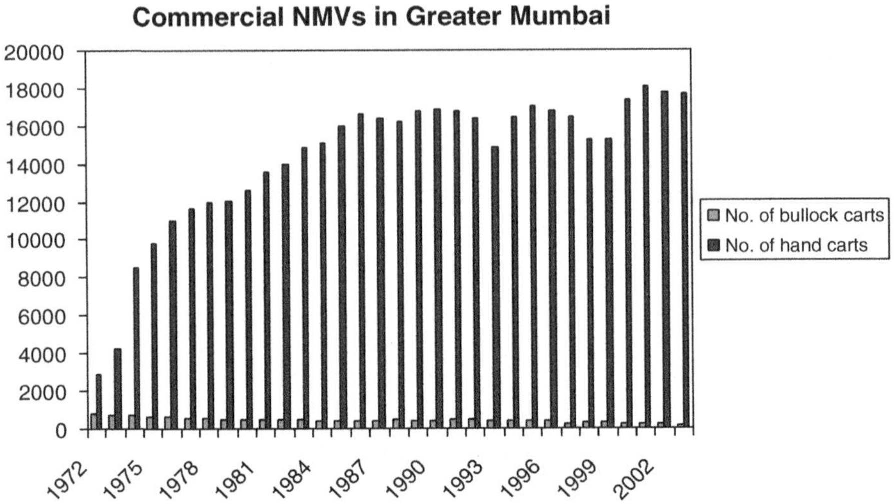

Fig. 5.1 Number of commercial non-motorized vehicles (NMVs) in Mumbai (Source: Prepared by the author with data from MMRDA 2005)

Mumbai. Apart from bicycle being used as personal transportation mode, NMT has been adapted as a cost-effective solution for trash disposal, agriculture produce transport, wholesale delivery, etc. Commercial applications dominate NMT uses in Mumbai. Figure 5.1 presents the number of commercial non-motorized vehicles (NMVs) in Mumbai. The number of commercial NMVs in Mumbai showed an increasing trend in spite of the growing congestion on roads (or it may be adding to it).

Mumbai has a significant number of road accidents every year with no significant decrease over time. This signifies the neglected infrastructure and facilities for NMT. Figure 5.2 presents the number of road accidents in Mumbai over a period of 30 years (1971–2003). About 30,000 people are meeting with accidents every year. As explained in Chap. 2, cyclists and walkers share road space with motorized traffic, and this puts NMT users at a very high risk.

Level of bicycle usage is linked to the developmental status of the place. In medium-sized cities in Japan, Germany, and the Netherlands, 40–60 % of all trips are made by walking and cycling, while in similarly sized cities in India, this share is as high as 80 % (Heierli 1993).

Though NMT users are the majority of the commuters in many places, they are often neglected in the design and modernization of transportation infrastructure. New construction and infrastructure upgrading projects often fail to provide physical infrastructure (e.g., overpasses or shoulders) for the NMT users, resulting in higher NMT-automobile accident rates, longer travel times for NMT users, or even a complete elimination of NMT traffic. The result of this negligence is a transportation system that in many ways favors cars and other motorized traffic to the detriment of the poorer segments of the population. Therefore consideration of

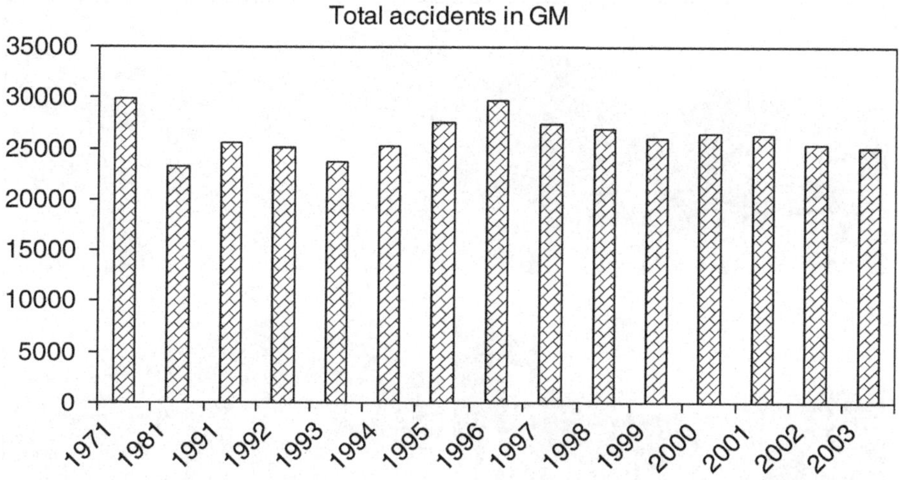

Fig. 5.2 Total number (annually) of road accidents in Mumbai (Source: Compiled by the author using data from multiple sources and Yedla 2007)

NMT during infrastructure design is an essential element to providing equitable transportation opportunities (Guitink et al. 1994; Guitink 1996).

In Indian metropolitan cities, with increasing urban sprawl and rising income levels, non-motorized transport modes tend to lose their importance. Share of bicycle trips out of the total trips in Delhi has declined from 17 % in 1981 to 7 % in 1994 (Mohan and Tiwari 1999). Increasing trip lengths, due to the urban sprawl, could be one major reason for this fall in share of non-motorized transportation apart from serious lack of infrastructure for NMT. The following plate presents the precarious situation of NMT users in one of the Indian cities (Plate 5.1).

5.2 Factors Favoring Non-motorized Transportation

Eighty percent of travel demand in Mumbai metropolitan area is met by the public transport, while the remaining 20 % is met by the personal modes of transportation like two-wheelers, cars, taxis, and auto rickshaws (motorized three-wheelers). The feeder trips to rail are mostly by bus. Commuting in Mumbai typically has three stages, viz., access leg,[1] primary peg,[2] and egress leg.[3] Access leg occupies 9.7 % of the total trip length, whereas primary leg averages at 85.3 % and egress leg at 5 %

[1] Access leg in commuting represents the distance covered between the origin and the access point where the passengers get into a transport mode covering the prime leg.

[2] Primary leg is the distance between the two major access points in commuting. It constitutes the major travel distance in daily commuting.

[3] Egress leg represents the distance covered between public access points to the destination.

Plate 5.1 Demonstration of model mix and risk to NMT users on road in one of the metros in India (Source: Pictures taken by the author)

(Rastogi 2002). Each of these components is met with different modes of transport. While the primary leg, the longest part of the journey, is typically met by either metro or bus, access leg and egress leg are shared among different modes. Figure 5.3 presents the share of different modes in access leg in Mumbai.

While cars and buses are marginal in meeting the access leg demand, walking, bicycles, two-wheelers, and auto rickshaws are dominating. Walking or combination of walking and bus to the access point of public transport is predominant. Forty-seven percent (47 %) of the access leg travel is still catered by motor vehicles, whereas only 7 % is met by bicycles. Average access leg in Mumbai is about 2.3 km (Rastogi 2002) with an overall average trip length at 15.06 km. It was reported in the literature that travel distance in the range of 3–5 km is ideal for cycling (GTZ 2005). Therefore, there is a scope for improving NMT in order to meet 100 % access leg in Mumbai.

The commuting distance in Mumbai extends up to 36 km in the north and up to 31 km in the east with transit stops at 1.6 km in GMR. The transverse distance from commuting corridors extends up to 5–6 km, thus providing an ideal setting for bicycle promotion and inclusion as an access mode to transit facilities. While NMT (particularly bicycles) is not very dominant in Mumbai, some of the Indian cities are still going well with non-motorized modes of transport. Figure 5.4 presents model split for major Indian cities.

However, the usage of NMT is inversely proportional to the size and economic status of the city. Kanpur which is relatively a smaller city has most of its trips met by NMVs. However, Mega cities like Delhi and Mumbai NMV share is not very

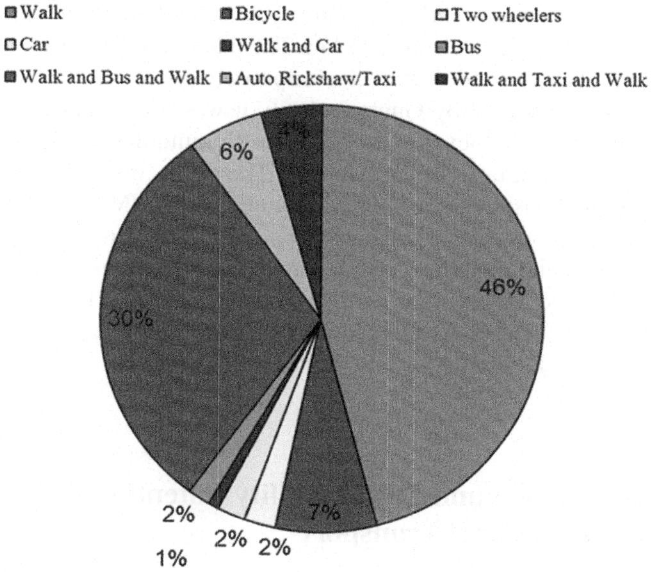

Fig. 5.3 Percentage share of various modes for access leg in Mumbai

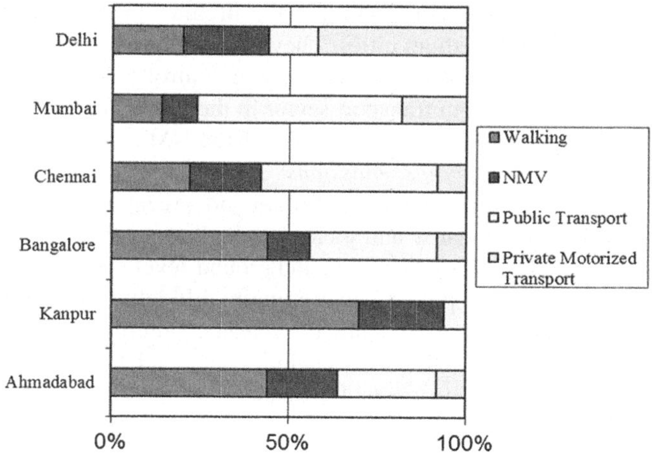

Fig. 5.4 Model split in major Indian cities (Source: Pendakur 2000)

significant. The size of city plays an important role in the usage of NMT and that is clearly seen in the usage of NMVs in Mumbai and Delhi. Delhi has a vast area and hence traditionally an NMT city. In contrast Mumbai is squeezed into a small area and the usage of NMV is very low, which could be due to heavy congestion. Efficient public transport system in Mumbai adds to this fact.

Mumbai, due to its extensive metro and bus network, generates substantial short-length trips. In a comprehensive study conducted by Rastogi (2002), it was observed that the maximum airline distance between a household and the nearest train station (in Vile Parle area) is 2,357 m. The maximum detouring distance was 4,020 m. It was also reported that 80 % of trips are work related in the study areas. In another study conducted by Gupta in 1986, it was found that the average trip length by walking or by bicycle was 5,100 m. Commuting in Mumbai typically follows access leg-primary leg-egress leg. The short trips covered by three-wheelers, taxis, and buses can be successfully tapped for NMVs in order to reduce load on public transport, reduce congestion, and also cut down on energy consumption and environmental emissions (particularly greenhouse gases).

The following sections present various advantages of NMT in social and economic terms other than environmental benefits and investigate possible reasons (barriers) restricting the implementation of NMT in Mumbai.

5.3 Socioeconomic and Sustainability Potential of Non-motorized Transport

5.3.1 Why Non-motorized Transport?

Non-motorized transport (NMT) modes are the lifelines for the transportation in rural areas and small and medium cities. They also provide benefits like reduction in congestion, energy-efficient transportation, and control on environmental emissions and greenhouse gas from transport sector in the urban centers (Tiwari 2003). The following paragraphs explain such benefits from NMT.

Control over greenhouse gas emissions: Pedestrians, bicyclists, and cycle-rickshaw passengers generate no air pollution, no greenhouse gases, and little noise. While emission standards and cleaner vehicles can greatly reduce certain emissions, reducing CO_2, nitrogen oxides, and ground-level ozone through tailpipe-focused measures alone is proving exceedingly difficult. These emissions are growing rapidly with motorization; while the "cleaner" alternatives provide a slight dip in the emissions curves, they, in the long run, tend to continue with such emissions. Even the switching to public transport cannot give away with GHG emissions, though it is relatively better than personalized motorized transport modes. Thus NMTs provide much-needed platform for the control of GHG emissions from transport sector.

Control over congestion: With the Governments making motorization-friendly policies jeopardizing the prospects of NMT, modern urban transport is facing tremendous levels of congestion. While fully occupied public transit vehicles are the most efficient users of the road space, bicyclists use less than a third of road space used by the private motor vehicles. Pedestrians use less than a sixth. Further, bicycles need 15 times less parking space compared to cars along the roadside (GTZ 2005).

Table 5.1 Share of different modes of transport in short-length trips between 1 and 3 km

	Share (%) of short trips between 1 and 3 km		
	Motorized modes	Bicycles	Walking
Surabaya	60	10	30
Germany	15	30	55
Mumbai (2000)	40–50	5–7	45–53

Source: Compiled using data from GTZ (2005) and Rastogi (2002)

Best transport mode for short trips: Though it depends on the geography of the place, in most of the developing country cities, more than 50 % of trips are of short length. A decentralized city like Delhi and Bangkok would have that many more short trips. However, a city like Mumbai which spreads in one direction would also have dominant short trips (as a part of the travel for work which has both long and short trips) as the metro-based rapid transit system involves short-length trips from home to the nearby access point to the public transportation mode. It essentially follows multi-model transit like walk and ride and ride and walk.

While the well-planned German cities have over 80 % of their trips very short, Surabaya (Indonesia) which is only 15 km across makes it possible, in principle, to make all trips by NMT (GTZ 2005). Mean trip length in Delhi was reported to be 5.1 km, whereas in an Indonesian city it was 3.3 km. In Chinese cities like Ningbo, it is as high as 6 km (Rastogi 2002). In 1998, Bagota had 70 % of its trips under 3 km covered by car, which otherwise could be the ideal length for bicycles. Even Asian cities whose per capita income is less than one-twelfth of Germany have 60 % of its short-length trips (less than 3 km) covered by motor vehicles. Table 5.1 shows such statistics.

Anantharajan (1980) reported that three-quarters of trips by slum dwellers in Chennai were made by walking, while one-third was covered by a distance of 1 km for journey to work and around 60 % covered a distance up to 3 km by walking.

Flexible: NMT provides a very flexible solution to accessibility, particularly when the resources are scarce. This is also true from the perspective of local service providers and policy makers (Guitink et al. 1994). People living in poverty face a wide variety of problems related to mobility, such as access to employment, social services and activities, educational opportunities, and household chores. NMT is a multipurpose tool that can be used for the door-to-door transport of persons and goods with improved travel time and route options.

Affordable: While buses may be affordable at the beginning of the month, fares often become prohibitive with dwindling cash over the following weeks, and people are left with no choice but walking. Affordability of NMT is a function of purchase price parity in relation to income. Operational costs rarely are a constraint with NMT. However, high retail prices frequently suppress the potential demand for NMT, such as bicycles, carts, and spare parts. Tables 5.2 and 5.3 present the details of comparative prices of different modes of transportation. As explained in the tables, non-motorized vehicles are far less expensive and maintenance-free compared to all motorized vehicles.

Table 5.2 Average prices of new vehicles from different modes of transport in few Asian cities (Unit: 1992$)

City	Bicycle	Cycle rickshaw	Motorized rickshaw	Motorcycle	Car
Phnom Penh	40	61	–	1,690	25,100
Dhaka	61	181	–	–	–
Kanpur	53	128	2,500	1,200	6,400
Surabaya	138	150	–	1,480	24,600
Manila	176	790	1,880	1,760	31,300
Chiang Mai	178	790	5,940	1,520	19,800

Source: World Bank (1995)

Table 5.3 Average annual operating and maintenance costs for different modes of transport in few Asian Cities (Unit: 1992$)

City	Bicycle	Cycle rickshaw	Motorcycle	Car
Phnom Penh	3	5	174	600
Kanpur	15	85	349	1,000
Surabaya	20	30	183	820
Manila	16	31	147	1,130
Chiang Mai	16	32	239	1,280

Source: World Bank (1995)

NMT for sustainability: NMT provides much-needed transportation at an affordable price for the low-income groups (urban poor). With the increasing urban sprawl and the travel distance for work, NMT has the potential to cater for the travel needs of the urban poor. While the end-of-the-pipe approach provides relief with certain air pollutants from automobiles, NMT presents a possible long-term solution for the greenhouse gas emissions, particularly from urban transportation systems. Therefore, while the smaller cities are encouraged to continue their NMT usage, the metropolitan cities should consider integrating NMT into their long-term transportation planning.

NMT provides other benefits like employment generation, savings in land requirements, and so on which are presented in the following sections.

5.3.2 Socioeconomic Benefits of NMT

Bicycles, as well as other forms of non-motorized transport, can bring significant economic, social, and environmental benefits. In the central business districts of many developing cities, congestion is a growing concern, affecting the health of inhabitants and economic activity. For instance, in Bangkok, an estimated $1.4 million worth of fuel is wasted everyday by vehicles idling in traffic. Adding to these economic costs are losses of revenue caused by the hampered ability of businesses to deliver goods and services and increased employee commuting times. A study by the Bangkok Metropolitan Administration stated that, had the person-hours lost from the average 44 working days spent in traffic each year been put to productive use, the gross national product would have grown by another 10 % (Pandakur 2000).

NMT provides substantial savings in infrastructure costs. Provision, operation, and maintenance of motorized facilities require resources ten times that of NMT facilities and five to six times that of public transport facilities (Litman 1998; Pettinga 1996; Pendakur et al. 1995; Sharma 1986). In a complete overview comparing two transit access systems, Replogle (1984) reported that the bike and ride system is a far more cost-effective strategy, related to energy conservation and pollution reduction, as compared to the park and ride system. The typical construction cost of a park and ride lot was reported as $3,640 per automobile space, whereas the capital cost of secure bicycle parking was reported as $50–500 per space, both excluding the land cost. Although a typical unattended park and ride lot costs $150 or more per year for operation and maintenance, this figure ranged from a few dollars to about $70 per year for bicycle parking.

NMT provides substantial savings in land requirements not only by means of less demand for road space but also land required for parking. Automobile parking typically requires as much as 330 ft^2 of land per space as compared to 6–12 ft^2 needed for ground-level bicycle storage spaces. And the bicycles can be parked in tier system which makes it that much more land efficient.

NMT provides direct economic benefits also such as savings in gasoline consumption, control over work loss due to congestion, etc. It was reported in the Chicago Area Transportation Study that the installation of secure bicycle parking at rail stations would reduce hydrocarbon emissions at a public cost of $311/t compared with $96,415/t for an express park and ride service. Another study reports that for each American park and ride commuter diverted to bike and ride travel, the gasoline use reduces by an average of 75 gal (285 l) per year. Another analysis reveals that by diverting automobile commuters to bike and ride travel, the average savings amounts to 400 gal (1,500 l) of gasoline each year for every new bike and ride commuter Replogle (1991).

NMT has tremendous congestion reduction potential and related economic benefits. In the USA, the congestion cost was reported to be varying between $100 billion and $300 billion (1990 price) per year which could be controlled by means of NMT (Goodland 1994; Komanoff et al. 1993). The same is reported as equal to 10 % increase in GDP for Thailand (Zegras and Birk 1994) and 240 working days per annum for India (Sharma 1986). By providing improved NMT facilities, the accidents can be controlled which otherwise costs $363 billion per year for the USA (1990 prices) and Rs. 25.125 million (1984 prices) for India (Gupta 1986).

Employment generation: Non-motorized transport, particularly bicycles and tricycles, provides employment opportunities at different stages, viz., manufacturing, servicing, and repair of the NMT fleet. The employment generation from NMT mostly belongs to informal sector. In Dhaka, Bangladesh, about 380,000 people are directly employed as rickshaw pullers where another 80,000 are in ancillary services, which together accounts for one-fourth of total employment in the Dhaka metropolitan area. This picture is much more staggering at the national level with one million people in direct services and another 250,000 people in ancillary services, making it 3.5 % of the national labor force. NMT is particularly vital in providing employment for unskilled low-income workers. In Indian cities like Patna, public transport along with NMT offers major source of livelihood (Replogle 1991).

Employment generation potential per unit of investment is higher in NMT. An investment of 100,000 Indian rupees in a conventional bus system in Patna produces two direct jobs, whereas the same amount produces six direct jobs in auto-rickshaw system and 75 direct jobs in cycle-rickshaw transport system. This indicates employment generation potential of NMT, which can be successfully utilized in the context of poverty reduction in Asian urban centers.

The concept of environmentally sustainable development implies that mobility solutions must encompass new targets, such as reducing energy consumption, improving public health and safety, and controlling environmental emissions including greenhouse gases. NMT provides the much-needed degree of freedom in achieving sustainability in transportation. These requirements of sustainability have to reflect on transport and traffic policy, especially in urban areas, and international funding agencies like the World Bank, Sida, and other UN bodies have to play their part in order to make it possible.

References

Anantharajan T (1980) Travel behavior of slum dwellers in Madras. In: Proceedings of the world conference on transportation research, London, pp 1634–1643

Cropper M, Bhattacharya S (2007) Public transport subsidy and affordability in Mumbai, India. Policy Research working paper # 4395. The World Bank, Washington, DC

Goodland RJA (1994) Urgent need for environmental sustainability in land transport in developing countries: an informal personal view. Transp Res Rec 1441:44–52

GTZ (2005) Training course on non-motorized transport, commissioned by Federal Ministry for Economic Cooperation and Development. Germany

Guitink P (1996) Strategic planning for non-motorized mobility, infrastructure notes. Transport No. OT-4. The World Bank, Washington, DC

Guitink P, Holste S, Lebo J (1994) Non-motorized transport: confronting poverty through affordable mobility. Infrastructure notes. Transport No. UT-4. The World Bank, Washington, DC

Gupta RG (1986) Delhi 2010 AD: cycle – an important mode even after the 20th century. In: Proceedings of the international conference on transportation system studies. IIT Delhi, India, pp 625–632

Heierli U (1993) Environmental limits to motorization. Neidermann, A.G, Switzerland

Komanoff C, Roelofs C, Orcutt J, Ketcham B (1993) Environmental benefits of bicycling and walking in United States. Transp Res Rec 1405:7–12

Litman T (1998) Potential transportation demand management strategies. Report of Victoria Transport Policy Institute, Canada. www.vtpi.org

Mohan D, Tiwari, G (1999) Sustainable transport systems: linkage between environmental issues, public transport, non-motorized transport and safety, economic and political weekly, June 1999

Mumbai Metropolitan Region Development Authority (MMRDA) (2005) Basic transport & communications statistics for Mumbai Metropolitan Region. Mumbai Metropolitan Region Development Authority, Mumbai, India

Pendakur SV (2000) A policy perspective for sustainable cities – non-motorized transport (NMT) in Asia, urban environmental management – outreach series. CUC-AIT, Bangkok

Pendakur VS, Badami MG, Lin YR (1995) Non-motorized transportation equivalents in urban transport planning. Transp Res Rec 1487:49–55

Pettinga A (1996) Quicker by bicycle – policy manual for bicycle friendly infrastructure. In report – non-motorized transport, The World Bank and the Inter-American Development Bank, pp 85–105

Rastogi Rajat (2002) A policy sensitive behavioural model of transit access. PhD thesis, Department of Civil Engineering, Indian Institute of Technology, Bombay, India

Replogle M (1984) Role of bicycle in public transportation access. Transp Res Rec 959:55–62

Replogle M (1991) Non-motorized vehicles in Asian Cities. World Bank Technical report 162, - Washington, DC

Sharma SK (1986) Bicycle renaissance. In: Proceedings of the international conference on transportation systems studies, IIT Delhi, pp 470–474

The World Bank (1995) Non-motorized vehicles in ten Asian cities- trends, issues and policies, report TWU20. World Bank, Washington, DC

Tiwari G (2003) Towards a sustainable urban transport system: planning for non-motorized vehicles in cities, transportation research and injury prevention programme. Indian Institute of Technology, Delhi

Yedla S (2007) Non-motorized modes of transportation for sustainable mobility: strategies for its adaptation in Mumbai, India, book on Air pollution control in the transportation sector: third phase research report of urban environmental management project. Institute for Global Environmental Strategies, Tokyo, pp 21–82

Zegras PC, Birk ML (1994) Moving towards integrated transportation planning: energy, environment and mobility in ten Asian cities. Transp Res Rec 1441:84–92

Chapter 6
Barriers for the Implementation of NMT and Other Cleaner Modes of Transport

6.1 Non-motorized Modes of Transportation

Chapters 4 and 5 presented various alternatives available in transportation sector in its pursuit to environment and social sustainability. Any effort to introduce new alternatives faces certain types of frictions in the system. Those frictions popularly termed as barriers can be broadly classified as administrative, financial, technical, social, cultural, and infrastructure barriers. It is essential to identify the right set of barriers so that necessary policy alternatives and measures can be devised to overcome those barriers. This chapter attempts to discuss all such barriers for the NMT and other cleaner options, and their prioritization is presented as a case study using analytic hierarchy process, which is gaining global popularity as a decision-making tool.

6.1.1 Social Barriers

Safety of the bicycle rider is recognized as a major barrier all around the world (Tiwari 2003; GTZ 2005). A study by the author (Yedla 2007) also confirms it to be an important issue controlling the usage of bicycles in Mumbai. In the absence of special lanes or sidewalks as in the case of Mumbai, bicycle riders share the road space with all other motorized modes and that puts them at a very high risk.

It is a generally believed that NMT (bicycles and walking) is a *slow-moving mode of transport* (World Bank 1995). This is the reason for the existence of motor vehicles—to speed up the movements. However, the opinion survey conducted reveals that slow motion is not a reason for not using NMT in Indian cities such as Mumbai. Instead, walking is preferred due to the fact that the approach roads are very congested and public and feeder (motorized three-wheelers) modes of

S. Yedla, *Urban Transportation and the Environment*,
DOI 10.1007/978-81-322-2313-9_6

transport are far slower compared to that of walking. Thus slower movement is not a barrier for the improvement in NMT usage.

Problem of theft is being considered as a major barrier for the implementation of bicycles in different parts of the world (GTZ 2005; World Bank 1995). However it is not perceived to be so in Indian cities. A survey conducted by the author revealed that in Mumbai theft is not seen as a major issue controlling the usage of NMT (bicycles). This could be a possibility that extensive usage of bicycles is not experienced in Mumbai and hence the respondents could not make a proper judgment. However, given the nature of extensive MRTS and fast-moving lifestyles in Mumbai, it is likely that theft may not be a problem in enhancing bicycles usage.

Like many other cities, *poor social acceptability of bicycles* is a major barrier in implementing bicycles in Mumbai. Started to increase the speed of movement, motorized vehicles have become a status symbol in cities all over the world. This is particularly so in developing countries. Thus bicycle is seen as a poor man's domain and that, to a great extent, limits the penetration of bicycle usage. This problem is universal but the solutions are not (World Bank 1995; Pendakur 2000).

As a constituent of NMT, walking provides a contrasting opinion. Walking is not limited to the poor and is equally used by all classes in the society. While it is a fact that other than poor also walk short distances in urban centers, the defining line for "poor" in Indian cities is tricky with many tax payers living in slums. Social dimension needs to be examined carefully in order to understand the complete dynamics of it and its linkage to the implementation of NMT.

Political unwillingness is not seen as a problem while the *attitude of both motor vehicle riders and the law enforcers* (police) is an important barrier. This essentially links to the lack of proper regulations in favor of NMT.

As city dwellers are generally aware of the benefits of NMT, *lack of awareness* does not qualify as a barrier. Non-use of NMT is more linked to social consciousness and lack of safety and infrastructure than lack of awareness. However, *lack of promotional campaigns* could be jeopardizing the effective implementation of NMT in Indian cities.

6.1.2 Financial Barriers

Financial barriers are the usual bottlenecks in the case of alternative technologies involving considerable investment, either capital or maintenance. Relevance of this type of barrier for NMT is only for bicycles and man-pulled carts or tricycles. *Affordability* is one of the major barriers for the promotion of NMT. However, in contrary to the popular belief that lack of affordability to buy a bicycle is a barrier, it is observed in Indian metropolitan cities that the higher affordability to buy automobile affects the choice of NMT negatively. Given the fact that bicycle is not socially accepted, ownership of motor vehicle replaces bicycle. This presents a case of "negative effect" of affordability.

However, the affordability of the poor needs a careful assessment. At this juncture again, it is necessary to bring the point that the definition of poor in metropolitan cities is typically different from the rest of the country. Though *lack of financial mechanisms to support the poor to buy a bicycle* is not considered as a major barrier, it is essential to further examine the case with "poor urban dwellers" before finally concluding on it.

6.1.3 Infrastructure Barriers

Lack of infrastructure has been regarded as a global bottleneck in implementing NMT (GTZ 2005; Replogle 1991; World Bank 1995, 2002). Traditionally infrastructure planning and development follows the need and requirements of motor vehicles, ignoring the facilities required for non-motorized modes. This problem is even more acute in developing country cities like Dhaka, Mumbai, and Hyderabad where even pedestrians share the road space with the motor vehicles. The narrow walking pathways are occupied by small-scale vendors, making it practically impossible to use them. Thus, *lack of infrastructure* like separate lanes for bicycles (not even partially segregation), parking spaces at key shopping and transit points, pedestrian pathways, elevated or subway crossing, etc., is a major barrier for the implementation of NMT in cities.

Though city dwellers have access to good-quality bicycles, rendering *lack of quality bicycles* not as a barrier, it is important to make available good bicycles at lower prices (domestic production?). Such a promotion improves the penetration of NMT in urban centers.

6.1.4 Institutional Barriers

Institutionalization of NMT use can be considered at local administration level as well as at national level. As issues related to bicycles deal with the majority of the public, it is necessary to institutionalize the process of integration into the mainstream transportation network (World Bank 2002). *Lack of institutional arrangement to integrate NMT in transportation planning at city/municipality level* is considered as a major barrier in Indian cities. At present there are no arrangements to integrate NMT in the planning and development of transportation infrastructure in Indian cities. Though various planning document such as the National Urban Transport Policy (NUTP) and Mumbai Metropolitan Region Development Authority (MMRDA) spelled out the need for development of NMT, there are no clear measures of integrating it into the mainstream transport network. Even the recently concluded projects such as the Mumbai Urban Transportation Project (MUTP) and Mumbai Urban Infrastructure Project (MUIP) and ongoing JNNURM initiatives do not consider the integration of NMT with the other transportation infrastructure (MMRDA 2011; MoUD 2007).

Lack of integration of NMT in national transportation plans and frameworks is another major barrier for the promotion of NMT in Indian cities. The present national transportation plan (NUTP) does mention about the promotion of NMT. However, it is not explicit on providing much-needed framework for such integration. Most of the success stories of NMT are due to the fact that they have been integrated in their respective national plans (World Bank 2002; GTZ 2005). Institutionalization at national level is necessary because of the fact that funds for the transportation infrastructure developments are allocated at national level (though PPP models are increasingly becoming popular at present).

6.1.5 Administrative Barriers

Non-motorized modes of transportation constitute a major element in development of mass rapid transit system (MRTS). In order to be successful, the public transport system needs to be complemented by NMT. This fact is amply supported by the fact that 40–60 % of daily trips in Indian metropolitan cities are made by walking. In contrary, in spite the fact that Mumbai is heavily dependent on its MRTS, NMT is not properly administered. The survey conducted by the author reveals that *lack of proper regulation for cyclists and their rights* and *lack of legal basis for NMT traffic management* are among the major barriers for the promotion of NMT (Yedla 2007). These aspects compliment with the institutionalization of NMT integration into mainstream transportation planning.

6.1.6 Natural/Culture Barriers

As NMT involves physical hardship, it is important to consider the local conditions under which NMT is used. As a general rule, tropical countries are not as suited for cycling as it is in temperate countries. Cities like Bangkok and Mumbai where the humidity levels are high and also have considerable resuspended dust are not a perfect fit for bicycle riders. Severe monsoon and hilly terrains are certainly not so conductive for NMT as well. However, it is interesting to note that the survey conducted by the author in Mumbai revealed that *climatic conditions like high humidity and monsoon* are not among the major barriers for the promotion of NMT. Such observations need further validation as it involves substantial personal bias in response.

Another study by Rastogi (2002) reveals that majority (24.14 %) of respondents who do not use walking or bicycle mentioned that they do not like to use NMT. Among the other barriers, difficulty in carrying baggage, heavy sweating, dust, and health control the usage of NMT. While only 7.5 % of respondents to the survey claim that they can afford automobile trip, 10 % say that availability of automobile trip is the reason for them not using NMT.

Based on the opinion survey conducted in Mumbai and other cities in India, the following are the major list of barriers for the promotion of NMT in Indian cities:

- Lack of Safety of the bicycle rider
- Poor social acceptability of bicycles
- Attitude of both motor vehicle riders and the law enforcers (police)
- Lack of promotional campaigns
- Higher affordability
- Lack of infrastructure
- Lack of institutional arrangement to integrate NMT in transportation planning at city/municipality level
- Lack of integration of NMT in national transportation plans
- Lack of proper regulation for cyclists and their rights
- Lack of legal basis for NMT traffic management

It is important for the city administrators in consultation/cooperation with the provincial and federal governments to address these barriers in order to have a better implementation possibility for non-motorized modes of transport in their respective cities. The following section identifies similar barriers for the promotion of cleaner modes of automotive transport options.

6.2 Cleaner Fuels and Technological Alternatives

The matrix of cleaner transportation option as presented in Chap. 4 includes cleaner fuel-based options such as CNG, electric vehicles, and biodiesel, alternative vehicle technology like electric vehicles and hybrid engine technology, cleaner technology such as four-stroke engines, and inspection, operation maintenance, and management measures. In line with the background and nature of resistance from within, as presented in the previous section, the following barriers are the limiting factors in implementing the cleaner alternative option of transportation. Such barriers are clustered against the category of alternative measure.

Barrier for cleaner fuels:	Availability of efficient technology/conversion kits
	Additional cost
	Inadequate resources and infrastructure
	Absence of enforcing mechanism

The above barriers represent technological, financial, resource and infrastructure, and awareness barriers, respectively. Compressed natural gas options, viz., bus, car, and three-wheelers, would have few common barriers. However, intensity of barrier depends on the alternative option chosen. For instance, conversion kits for CNG cars are not indigenously made and importing them puts a lot more burden on the users. Unlike buses, where the burden on passengers is marginal, cars are individual units and the additional cost of conversion kit puts stronger burden on the owner.

In the initial stages of conversion, problems with the retrofits give rise to serious safety issues. Thus, availability of efficient retrofits and their easy accessibility would be an important barrier for the greater penetration of CNG cars. Additional cost is another very important barrier for CNG and other alternative-fueled cars. Retrofitting kits are imported and hence are considerably expensive. Dedicated CNG vehicles from the standard car manufacturers come with a premium (IGIDR 2004). The additional cost on cleaner vehicles acts as a finance barrier for the implementation of this cleaner option. Bringing down the difference between the prices of conventional and CNG vehicles facilitates in removing this barrier.

Infrastructure and resource availability are of great importance for CNG technology and other alternatives. Unlike for CNG buses, infrastructure is very important barrier for CNG cars. Due to lack of widespread network of filling stations, CNG car owners face problems and that makes this option not so well sought after. Removing this barrier makes a way for smooth implementation of this alternative transportation option (TERI 2002; World Bank 2004). Lack of awareness as a barrier follows a similar trend. Cars, having a character of individual ownership, need much more awareness campaign compared to that for buses. Lack of awareness about the benefits of CNG cars over the conventional cars is a major reason for people not opting for this energy-efficient and cleaner mode of transport (IGIDR 2004).

Three-wheelers are used as means of public transport though it is a kind of personalized transportation mode. As a feeder service mode, three-wheelers present a special case, which stands midway between buses and cars (Pucher et al. 2007; Yedla 2005; United Nations 2001). Hence, this mode presents a very special scenario for all these barriers identified. Technology availability may not have similar impact on CNG three-wheelers and CNG cars due to the difference in ownership. But the same barrier may show much more indirect influence by means of reduced cost due to easy availability of technology. Factors like safety and increase in noise levels may not influence the implementation of CNG three-wheelers.

Additional cost as a barrier is very important due to the reason that it not only influences the owner of the vehicle but also the user. Though the operating cost is less in the case of CNG three-wheelers, the higher capital cost creates some friction for its implementation. Though automobile makers bring out CNG three-wheelers, high price plays a hindrance. Financing agencies have a major role to play in removing this barrier. Argument about the other two barriers, viz., lack of resources and infrastructure and lack of enforcing mechanism, remains similar to that of CNG cars. Delhi in the initial years of CNG implementation has witnessed a severe problem of scarce CNG filling stations with CNG three-wheelers lining up for a kilometer or two, waiting for hours together in order to fill their fuel tanks (TERI 2002; World Bank 2004; Yedla 2005). This can potentially affect their incomes and hence a prominent barrier. Though authorities view this as a short-term problem, it could pose a long-term problem in terms of availability of CNG. Another dimension to this problem is the leakage at CNG filling stations, which adds to the GHG emissions. Lack of proper and efficient filling technology leads to such externalities and acts as one of the barriers for the implementation of this cleaner fuel option.

Following are the barriers for the alternative technologies such as electric vehicles, hybrid technology, hydrogen power vehicles, etc.

Technology	Barriers
Alternative vehicle technology (such as electric vehicle)	Availability of alternative technology
	Additional cost
	Lack of awareness
	Absence of mechanism for its implementation

Barriers for this alternative option represent technological, financial, awareness, and administrative barriers, respectively. Battery-operated vehicles are still under examination for various facts, and hence the specifics of technological parameters are yet to be determined (Delucchi 1993; United Nations 2001; Delucchi and Lipman 2001; Lave and MacLean 2002). Hence, technology availability as a barrier is of immense importance. Though the technology has been proven in many places, it is yet to reach the developing world and hence is a major barrier for the implementation of this option (United Nations 2001). Additional cost may not be a stronger barrier as its maintenance costs are negligible, making it an economically viable option. However, the initial investment is a barrier. This alternative option has very distinct advantages like negligible maintenance cost, low operating cost, and absolutely emission-free ride. To achieve a better penetration, it is important to have a strong awareness campaign for these cleaner alternatives.

6.3 Prioritization of Alternatives by Analytic Hierarchy Process

Not just the barriers and their nature but their importance should also be considered while assessing them. Ranking of barriers with due consideration to all factors (criteria) and actors involved should be done in order to identify the most important barriers. This would help in making policy suggestions and steps to alleviate these barriers. Barriers that are identified and analyzed in the previous sections are ranked in a multi-criteria framework adopting analytic hierarchy process (AHP).

6.3.1 Analytic Hierarchy Process

Priority setting exercise is well established with considerable literature support to it. Usually ranking is done using quantitative approach with various optimization techniques. There are many methodologies documented even for qualitative technique-based priority setting. However, the problem takes a different shape when it involves group aggregation. For priority setting from a group of members and based on qualitative and/or quantitative criteria, analytic hierarchy process (AHP) was found to be one of the most widely applied methods (Yedla 2001). The AHP model is a suitable tool for this because it combines deductive approach and systems approach of solving problems into one, integrated logical framework (Ramanathan and Ganesh 1995). The three principles of guidance in AHP are decomposition, comparative judgment, and synthesis of priorities (Saaty 1990). This method is simple, practical, systematic, and effective. It has been widely used for conducting analysis of energy planning, distribution of resource planning, conflict resolution, selection of projects, etc., and useful results have been obtained (Kaundinya and Ramanathan 1994; Ramanathan 2000; Yedla 2001).

Analytic hierarchy process was developed by Thomas L. Saaty in the 1980s. It is one of those specialized tools based on priority theory that are applied for decision-making process both for individual decision-making and group decision-making. AHP, unlike other decision-making tools, has the capability of handling both qualitative and quantitative parameters in its process. An integrated multi-criteria decision-making system with other methods, viz., goal programming, linear programming, etc., is also proposed (Ramanathan and Ganesh 1995). Unlike quantitative criteria, qualitative criteria are difficult to judge and AHP provides a platform for such situations. Figure 6.1 shows the general principle of AHP schematically. The fundamental principle of AHP is the "pair-wise comparison of different variables which are given numerical values for their subjective judgments on relative

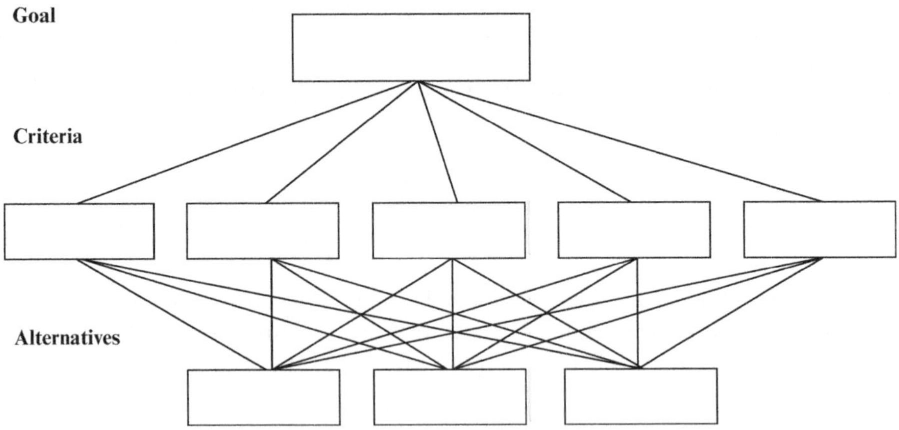

Fig. 6.1 General principle of AHP (Source: Saaty 1989)

Table 6.1 Pair-wise comparison scale for AHP preferences

Verbal judgment of ranking	Numerical rating	Explanation
Equal importance	1	Two activities contribute equally to the objective
Moderate importance of one over the other	3	Experience and judgment slightly favor one activity over another
Essential or strong importance	5	Experience and judgment strongly favor on activity over another
Very strong importance	7	An activity is strongly favored and its dominance demonstrated in practice
Extreme importance	9	The evidence favoring one activity over another is of the highest possible order of affirmation
Intermediate values between the two adjacent judgments	2,4,6,8	When compromise is needed

Source: Saaty (1989)

importance of each of the variable following a hierarchy that lead to assigning relative weightages to those variables." This process breaks down a complex and unstructured situation into components, forming a hierarchy. Then the scale developed by Thomas L. Saaty (Table 6.1) is used to make pair-wise subjective comparisons which are subsequently synthesized to get the final priorities of the alternatives.

6.4 Application of AHP to Prioritize Barriers

This section demonstrates the application of AHP to prioritize a number of given alternatives. AHP is applied to prioritize barriers to the implementation of cleaner alternatives in Mumbai. A similar approach can be adopted to apply AHP to any other prioritization process in transport sector. This case study[1] also serves as a tutorial for the research applying AHP to transport sector.

The hierarchy system is formulated with objectives (termed as the focus), criteria, and the different options (the barriers). Information is collected by means of personal interviews with a specially designed questionnaire. The data collected is then used to rank the barriers for the alternative option of using alternative technology (electric vehicles). Data collected through questionnaire survey is analyzed using Expert Choice[2] to arrive at priorities.

[1] Parts of the material presented in this chapter are referred from Yedla (2003) and IGIDR (2001).

[2] Expert Choice is a software developed based on analytic hierarchy process (Expert Choice 2000).

6.4.1 Formulation of the Problem

6.4.1.1 Objective

The overall objective is to rank the barriers to implementation of battery-operated vehicles in Mumbai transport system.

6.4.1.2 Actors

Different actors have different perceptions for each barrier. The actors having influence over the ranking of barriers and included in the case study are energy/environment experts (EE), policy makers (PM), users (Users); manufacturers of equipment (Manufacturers), retailers of equipment (Retailers), and business community (investors/lenders) (BC).

6.4.1.3 Criteria

In order to rank the barriers, it is essential to define the criteria based on which the ranking of barriers is attempted. These criteria are based on the characteristics of barriers. While some barriers are easy to understand, some may be incidental. At times, barriers may be intentionally created to achieve certain other goals or to restrict the access. For the conditions prevailing in Mumbai and its transport system, the following are the criteria selected based on which the priorities for barrier is set:

Criteria 1: Monetary cost to remove the barrier
Criteria 2: Level of political/bureaucratic efforts
Criteria 3: Impact on adoption
Criteria 4: Life of barrier

Criteria are very important in setting the priorities of barriers. Pair-wise comparison differs to a great extent with the criteria based on which the comparison was made.

6.4.1.4 Decision Alternatives

Barriers to be ranked are the decision alternatives in the case study. The barriers to be ranked are those identified in the previous section.

1. Availability of alternative technology
2. Additional cost
3. Lack of awareness
4. Absence of mechanism for its implementation

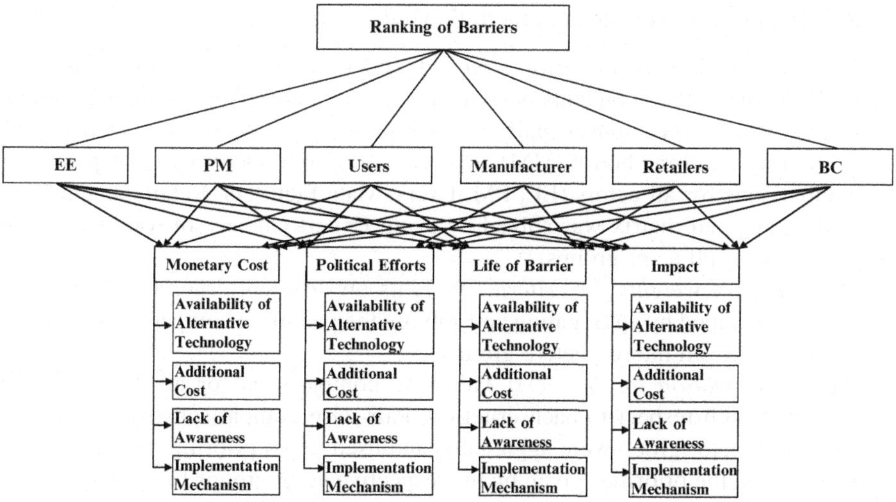

Fig. 6.2 AHP hierarchy tree for the ranking of barriers

Figure 6.2 shows the representative hierarchy tree employed in the case study. Priorities for different actors, criteria, and alternatives are established, and it is done based on pair-wise comparison scale shown in Table 6.1. As a precondition for AHP, the inconsistencies of all pair-wise judgment should be less than or equal to 10 % (0.1).

6.4.1.5 Pair-Wise Comparison Matrix

Pair-wise comparisons are fundamental building blocks of the AHP. Scale for pair-wise comparison matrix was provided by Saaty for quantifying the qualitative priorities given by the respondent. A specimen table showing pair-wise matrix is given here for six elements (6 × 6 matrix).

Factor/criteria	Barrier 1	Barrier 2	Barrier 3	Barrier 4	Barrier 5	Barrier 6
Barrier 1	1					
Barrier 2		1				
Barrier 3			1			
Barrier 4				1		
Barrier 5					1	
Barrier 6						1

Ranking of barriers is done in the following steps:

1. *Derivation of weights for actor groups*: Pair-wise comparison matrix was filled by all actors. All respondents have made pair-wise comparisons among various actors. All the 40 responses (pair-wise comparisons) are aggregated by adopting geometric mean method (GMM)[3] to arrive at a combined matrix of pair-wise comparisons for all actors (EE, PM, Users, Manufacturers, Retailers, and BC). The aggregated matrix was solved by using Expert Choice software to get the priorities for all actor groups.
2. *Derivation of weights for criteria*: Weights on different criteria by the group were determined by aggregating and solving the pair-wise comparisons made by individuals from the respective groups.
3. *Weight derivation for barriers*: Pair-wise comparisons for all barriers were collected with respect to each criterion. Pair-wise comparisons by individuals from each actor group were aggregated and used to get priorities for barriers by factoring in the priorities given by the respective actor groups for each criterion.

Weights assigned for different actors as well as criteria are used for the final ranking of barriers. The entire data set of the questionnaire survey has been distributed into different actor categories, and geometric mean method (GMM) was employed to arrive at a common pair-wise matrix, and weights of the barriers were derived using this matrix which is solved by means of *Expert Choice* software. With the derived weights for each actor group and criterion, aggregated group opinion was achieved on the ranking of barriers. As a common measure to meet the limit of inconsistency of AHP, responses of all individuals are tested for its inconsistency index and the questionnaires exceeding the limit were eliminated from the final set of responses.

6.4.1.6 Design of Questionnaire and Survey

A specially designed questionnaire was used to collect the pair-wise comparisons. The number of respondents from each actor group was kept uniform. Total sample size for Mumbai was 40 with almost equal representation from different sectors.[4] Actors interviewed include individuals from various ministries, viz., the Ministry of Environment and Forests, Ministry of Petroleum and Natural Gas, and Ministry of Urban Development; officers from the Central Pollution Control Board; experts from the Indian Institute of Technology, Bombay, and the Energy Research Institute, Delhi; representative members from the Confederation of Indian

[3] GMM is the most commonly used aggregation methodology adopted in AHP. More details can be obtained from Yedla (2001).

[4] AHP does not present any limitation on the sample size. The questionnaire can even be used with only one response. Multiple responses are aggregated to get a mean response representing the group of stakeholders. It would not be practical to have larger "equally represented" sample size with actors like policy makers, manufacturers, and experts.

Industry (CII), Society of Indian Automobile Manufacturers (SIAM), Center for Science and Environment (CSE), automobile research and testing agency, and automobile manufacturers; and transport service providers and maintenance agencies (IGIDR 2001).

6.4.2 Derivation of Rank for Barriers

Weights derived for all actor groups are presented in Fig. 6.3. Overall consensus of the group showed that user's opinion should be given maximum priority in ranking the barriers for the implementation of alternative options. Users are followed by energy and environmental experts (EE) and manufacturers (M). Maximum weight are given to opinion of users in handling the barriers which strongly supports the general feeling that most of advanced technological options fail due to the poor user (community) participation and inefficient strategic policy making and penetration (United Nations 2001; World Bank 2004). As the option considered is an advanced technological alternative option, the manufacturer's opinion gets more weight than the policy makers. These weights derived for different actors are used in ranking the set of barriers.

Any one-to-one comparison of barriers should be based on a particular criteria and comparison of any two barriers would differ based on different criteria. Some criteria show more importance than others in ranking a particular barrier. For instance, availability of technology as a barrier may gain more priority with reference to monetary cost (as criteria), but it may get less priority with reference to level of political efforts required to remove the barrier. Hence, all criteria have been subjected to pair-wise comparison, and their weights were derived in the procedure explained in the previous sections. Weights given by each actor group for different criteria are presented in Table 6.2.

Priorities derived for actor groups and criteria are used to prioritize barriers.

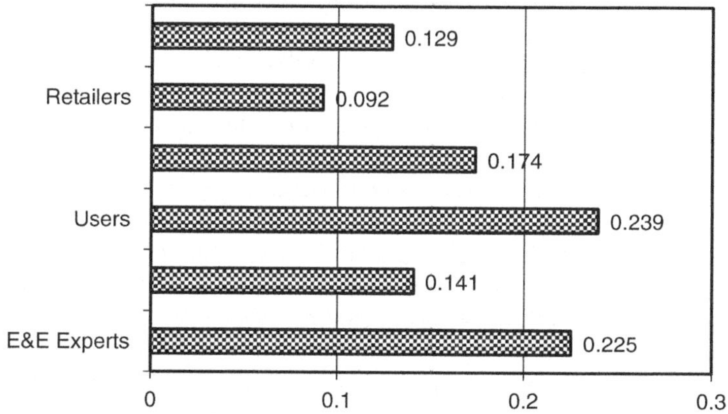

Fig. 6.3 Derived weights for various actors (after group aggregation)

Table 6.2 Weights for criteria

	Energy and environmental experts (0.226)	Policy makers (0.185)	Users (0.278)	Manufacturers (0.125)	Retailers (0.057)	Business community (0.129)
Monetary cost	0.137	0.132	0.238	0.142	0.078	0.195
Political/ bureaucratic efforts	0.132	0.134	0.413	0.218	0.446	0.499
Impact on adoption	0.465	0.492	0.161	0.353	0.417	0.141
Life of barrier	0.266	0.243	0.188	0.287	0.059	0.165

Values in parenthesis indicate the priorities given to the respective actors

6.4.2.1 Prioritization of Barriers

Alternative transportation option		Battery-operated three-wheelers
Criteria	Monetary cost	(C1)
	Level of political/bureaucratic efforts	(C2)
	Impact on adoption	(C3)
	Life of barrier	(C4)
Barriers (alternatives)	Availability of alternative technology/ conversion kits	(B1)
	Additional cost	(B2)
	Lack of awareness	(B3)
	Lack of enforcing mechanism	(B4)

For the implementation of BOV three-wheelers, additional cost (B2) gets more priority over the other three barriers. This may be due to the fact that people are not aware of the cost-effectiveness of this option and are concerned with the general feeling that any advanced technology would be expensive. Lack of technology availability as a barrier followed the additional cost barrier. Users gave almost equal weight to all barriers, which could reflect the poor awareness levels about this new alternative option. Lack of technological availability is a common concern, and unlike CNG options, most actors gave this barrier a higher ranking. Hence, lack of technology as barrier was chosen as the second most important barrier for the implementation of battery-operated three-wheelers.

It is interesting to note that lack of awareness did not gain much of importance compared to barriers like technology availability and cost. This may mean that once the technology is available at the least cost, awareness would not remain a barrier for the implementation of battery-operated three-wheelers.

Final ranking of barriers for the implementation of battery-operated three-wheelers was done with group aggregation and using the weights derived for actors and criteria. Aggregated group ranking of barriers for the option of BOV three-

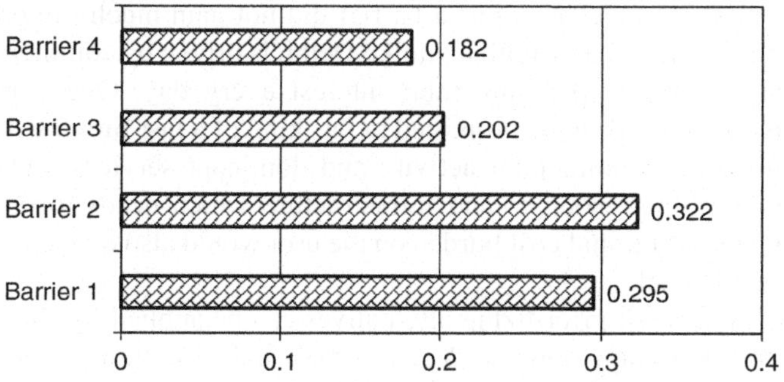

Fig. 6.4 Ranking of barriers for BOV three-wheelers implementation in Mumbai, given by the entire group (after aggregation)

Table 6.3 Barrier ranking for all three alternative options

	B1	B2	B3	B4
CNG car	0.128	0.172	0.505	0.195
CNG 3-W	0.263	0.262	0.288	0.187
BOV 3-W	0.295	0.322	0.202	0.182

wheelers is given in Fig. 6.4. Additional cost as a barrier dominates the priorities followed by availability of alternative technology. Hence, policies should be formulated and action needs to be taken to come over these barriers for the better implementation of BOV three-wheelers. Final ranking of barriers for the option of BOV three-wheelers is as follows:

B2 > B1 > B3 > B4 Additional cost > availability of alternative technology/conversion kits > lack of awareness > lack of enforcing mechanism

A similar prioritization exercise conducted for the other alternatives (cleaner fuel options) revealed a similar ranking of barriers for the respective options as presented in Table 6.3.

Measures to reduce additional cost barrier could be policies to subsidize the additional expenditure, policies to provide interest subsidies on loan to covert vehicles, financial incentives like free or priority parking and separate lanes for alternative fuel vehicles (AFVs), policies to provide free inspection and maintenance, policies to provide AFV access to the otherwise restricted areas, and devising tax on polluting vehicles and earmarking such revenue to provide subsidies for AFVs (Pucher et al. 2007; United Nations 2001; World Bank 2004). Thus the Government has to carefully set the financial incentives for cities based on their economic status in order to successfully implement cleaner and alternative fuels in Indian cities.

Lack of enforcing mechanism as a barrier did not gain much importance. This could be due to the fact that Mumbai, unlike Delhi, is driven by commercial activity with millions of people pursuing their interest every day. Once proved of its potential, no option needs to be promoted in this kind of a fast-moving metropolitan city with substantial commercial activity and dominant services and production sectors. It is only the resources and infrastructure which needs to be taken care of. Removing the additional cost burden on the user would also make the alternative options more adaptable.

A similar exercise of prioritizing alternatives can be applied for the selection of alternative transportation options (Yedla 2003) and also transportation policies aimed at implementing cleaner options (IGIDR 2004).

References

Delucchi MA (1993) Greenhouse gas emissions from the use of new fuels for transportation and electricity in the developing world. Transp Res A 27A(3):187–191

Delucchi MA, Lipman TE (2001) An analysis of the retail and lifecycles cost of battery-powered electric vehicles. Transportation Research Part D 6:371–404

Expert Choice (2000) Software manual

GTZ (2005) Training course on non-motorized transport, commissioned by Federal Ministry for Economic Cooperation and Development

IGIDR (2001) "Techno-economic assessment of alternative options in urban transport sector for Delhi and Mumbai" under the project of analysis of technological alternatives for mitigation of GHG emissions from urban transport sector in selected Asian cities. IGIDR, Mumbai

IGIDR (Indira Gandhi Institute of Development Research) (2004) Measures to overcome barriers for wider adoption of cleaner and energy efficient technologies in Delhi and Mumbai transport systems, Report submitted to AIT under. ARRPEEC, Thailand

Kaundinya S, Ramanathan R (1994) Incorporating qualitative objectives in integrated resources planning: application of analytic hierarchy process and compromise programming. IGIDR RP series 126. IGIDR, Mumbai

Lave LB, MacLean HL (2002) An environmental-economic evaluation of hybrid electric vehicles: Toyota's Prius vs. its conventional internal combustion engine corolla. Transportation Research Part D 7:155–162

Ministry of Urban Development (MoUD) (2007) National urban transportation policy. Ministry of Urban Development, Government of India

MMRDA (Mumbai Metropolitan Regional Development Authority) (2011) Regional plan for Mumbai metropolitan development area. MMRDA, Mumbai

Pendakur SV (2000) A policy perspective for sustainable cities – non-motorized transport (NMT) in Asia, urban environmental management – outreach series. CUC-AIT, Thailand

Pucher J, Peng Z, Mittal N, Zhu Y, Korattyswaroopam N (2007) Urban transport trends and policies in China and India: impacts of rapid economic growth. Transp Rev 27(4):379–410

Ramanathan R (2000) A holistic approach to compare energy efficiencies of different transport modes. Energy Policy 28:743–747

Ramanathan R, Ganesh LS (1995) Energy resource allocation incorporating quantitative and qualitative criteria: an integrated GP-AHP models. Soc Econ Plan Soc 29:197–218

Rastogi Rajat (2002) A policy sensitive behavioural model of transit access. PhD thesis, Department of Civil Engineering, Indian Institute of Technology, Bombay, India

Replogle M (1991) Non-motorized vehicles in Asian cities. World Bank technical report 162, - Washington, DC

Saaty RW (1989) The analytical hierarchy process – what it is and how it is used. Math Model 9 (3–5):161–176

Saaty TL (1990) The analytic hierarchy process: planning, priority setting, resource allocation. McGraw-Hill International, New York/London

The Energy Research Institute (TERI) (2002) Pricing and infrastructure costing for supply and distribution of CNG and ULSD to the transport sector, Mumbai, India. TERI Press, New Delhi

The World Bank (1995) Non-motorized vehicles in ten Asian cities- trends, issues and policies, report TWU20. World Bank, Washington, DC

The World Bank (2002) Cities on the move, a World Bank urban transport strategy review. World Bank, Washington, DC

The World Bank (2004) Reducing air pollution from urban transport. The World Bank, Washington, DC

Tiwari G (2003) Towards a sustainable urban transport system: planning for non-motorized vehicles in cities, transportation research and injury prevention programme. Indian Institute of Technology, Delhi

United Nations (2001) Review of developments in transport and communications in the ESCAP region 1996–2001. UN Economic and Social Commission for Asia and the Pacific (UNESCAP), Bangkok

Yedla S (2001) Application of analytic hierarchy process for the selection of alternative options for urban transportation in Delhi – analysis of group aggregation methodology adopted in AHP. Post-doctoral Fellowship Report. AIT, Bangkok

Yedla S (2003) Urban environmental evolution: the case of Mumbai. IGES

Yedla S (2005) In: Parikh KS, Radhakrishna R (eds) Urban transportation trends, alternatives and policy issues, India Development Report 2004-2005. Oxford University Press, New Delhi

Yedla S (2007) Non-motorized modes of transportation for sustainable mobility: strategies for its adaptation in Mumbai, India, book on air pollution control in the transportation sector: third phase research report of urban environmental management project. Institute for Global Environmental Strategies, Japan, pp 21–82

Yedla S, Shrestha RM (2003) Multi-criteria approach for the selection of alternative options for Delhi transport system. Transp Res A Policy Prac, Elsevier Publisher 37(8):717–729

Yedla S, Shrestha RM, Gabrial A (2005) Environmentally sustainable urban transportation – comparative analysis of local emission mitigation strategies vis-à-vis GHG mitigation strategies. Trans Policy, Elsevier 12(3):245–254

Chapter 7
Policy Analysis for Sustainability

Chapters 4 and 5 describe the possible alternative ways to include environmentally friendly elements into transportation planning so as to make the future transportation trends more sustainable and equitable. Chapter 6 identifies all possible barriers that the city administration may face while trying to plan transportation system more toward environmental sustainability. The present chapter identifies and explains the policy measures that can reduce or remove the barriers so that cleaner and efficient transport alternatives can be implemented.

The policy measures identified separately for the implementation of NMT and cleaner fuels and alternative technologies are applied to a case (study) to explain the process of prioritizing the same based on multi-criteria approach. While one case considers multi-stakeholder approach, the other case study considers supra user approach in getting the responses for the analysis. Prioritization of policy measures helps policy makers to choose an important and yet appropriate policy that has the potential to help in the implementation of the alternatives.

7.1 Policy Measures to Promote Non-motorized Modes of Transport

7.1.1 Identification of Policies Measures

Chapters 5 and 6 explain the importance of having non-motorized modes of transport (NMT) in order to achieve sustainable mobility in urban centers. This chapter attempts to analyze a set of identified policies/strategies to promote NMT and prioritize them based on multiple-criteria-based methodology. In the characterization of barriers as identified and analyzed in Chap. 6, the following policies

---

Parts of this chapter are extracted from Yedla (2007).

S. Yedla, *Urban Transportation and the Environment*,
DOI 10.1007/978-81-322-2313-9_7

are identified based on literature and can be employed to remove barriers and achieve better implementation of non-motorized modes of transport in Indian cities:

- Policies to integrate NMT with the public transport system planning at city/municipality level
- Formulation of National Strategy for NMT as a facilitating framework for local plans
- Policies to extend more specific incorporation of NMT in national transportation policy
- Incorporation of standards for the bicyclist and pedestrian provisions in new road infrastructure design
- Policies to provide NMT-friendly infrastructure (providing separate bicycle lanes on arterial roads, constructing wider pedestrian ways with cycling facility, providing subways and overhead passages at major intersections, and providing bicycle parking places at all public transport stations and public places)
- Policies to provide soft loans and relaxing excise duty on importing bicycles/bicycle parts
- Policies to make regulations safeguarding NMT users and to develop respect for the NMT users among the law-enforcing officers (police) and the motor vehicle users
- Policies to conduct awareness-raising campaigns

Any policy measure would have multiple affect other than the principle objective. While some of them could be positive impacts, there could be some negative externalities resulting from such policy measures. Therefore, it is important to have a comprehensive assessment of the proposed policy measures. Hence, a descriptive analysis of socioeconomic costs and benefits of the chosen policy alternatives and measures is presented in the following section.

7.1.2 Assessment of Socioeconomic and Environmental Benefits

Implementation of the proposed policies results in micro and macro impacts spanning over inter- and intra-sectors. This section presents such impacts separately for each of the policy measures identified.

7.1.2.1 Policies to Integrate NMT with the Public Transport System Planning at City/Municipality Level

Such integration demands additional budgetary allocations to meet the additional features and needs of NMT. Cities, with their limited resources, may need support from the Central Government and/or international financing agencies. Integrating NMT into the public transport system needs capacity building both for planners and the construction teams.

Such policies require substantiation of the existing administrative setup and technological capabilities at city level. Thus, it not only creates more jobs but also burdens the Government with additional budgetary requirements. By integrating NMT with the public transport system, one can achieve better results in terms of congestion reduction and environmental and public health benefits. Such integration policies may not burden the user unless the bicycle registration is made mandatory and also charged at parking places.

7.1.2.2 Formulation of National Strategy for NMT as a Facilitating Framework for Local Plans

This measure needs most of its efforts at national policy development. With a vast network of research institutes in the area of transportation planning and handful of experts, the development of such framework should not require much additional efforts. The National Government may need to constitute a committee to examine the possibility and develop such framework. As this would be a facilitating framework, it may not lead to additional economic burden on either National Government or Local Government. On firsthand criteria, it may not have considerable employment generation potential or any other social and environmental benefits. However, presence of such framework facilitates better adaptation of NMT at city level, and such efforts result in least costs.

7.1.2.3 Policies to Extend More Specific Incorporation of NMT in National Transportation Policy

This policy measure would result in more explicit NMT strategy at National Level which means more budgetary allocation at National level for transportation development and also financial supplies to the state and/or cities. Such National NMT strategy would result in more economic burden on the state/city administration. Such measures would result in substantial employment generation. Implementation of such strategies needs considerable capacity development activities, which adds to the burden on the Government. Users may also be subjected to additional burden if provisions are made for user fee as a mechanism of cost recovery. National NMT strategy would result in substantial environment and social benefits.

7.1.2.4 Incorporation of Standards for the Bicyclist and Pedestrian Provisions in New Road Infrastructure Design

This alternative measure needs capacity development for the technical man power at city level. As this is a measure at design level, its employment generation potential could be very low. It may not lead to a direct economic burden on either National or local Governments. But such incorporation of standards makes it easier to implement the NMT features in the overall transportation planning. Thus it has high potential in resulting environmental and social benefits. This further leads to development and growth of NMT industry wherein there is a possibility of increased employment.

7.1.2.5 Policies to Provide NMT-Friendly Infrastructure

This measure includes construction of separate bicycle lanes on arterial roads, construction of wider pedestrian ways with cycling facility, providing subways and overhead passages at major intersections, and providing bicycle parking places at all public transport stations and public places. This policy measure involves substantial economic burden on Local Government and National Government along with international aid. Construction of bicycle lanes costs substantially high and such projects in the developing countries need international support. Details on such costs are discussed in the earlier sections.

This policy measure has high employment generation potential. It was reported in the literature that investment of Rs. 100,000 in NMT would lead to 75 new direct jobs whereas it would give 2 jobs in public transport system and only 6 in the case of auto-rickshaws. Thus, this policy has potential to create jobs both in the construction stage and operational stage.

Such infrastructure development involves cost recovery process which may lead to user charges. However, the level of burden depends on the Government policy and less the burden on the user more on the Government.

Lack of infrastructure has been rated as one of the major barriers for NMT, and provision of such facilities would make headways toward sustainable transportation in terms of reduced congestion, energy conservation, and environmental benefits. However, such success depends highly on the well-designed awareness campaigns.

7.1.2.6 Policies to Provide Soft Loans and Relaxing Excise Duty on Importing Bicycles/Bicycle Parts

This would have a positive impact on bicycle and tricycle ownership. Though Mumbai enjoys higher per capita income compared to the other cities, commercial NMT vehicles need such financial incentives for their penetration. As explained in the earlier sections, intermodal transportation has great potential in commercial applications. This means some level of employment generation at commercial level NMT. This policy measure would burden the Government to an extent depending on what degree of subsidy is offered.

7.1.2.7 Policies to Make Regulations Safeguarding NMT Users and to Develop Respect for the NMT Users Among the Law-Enforcing Officers (Police) and the Motor Vehicle Users

This policy measure involves considerable levels of monitoring and hence would need additional man power and also burden the Government. This needs capacity building and awareness programs over the long term and hence burdens the Local Government. However, regulations have to be made alongside with the infrastructure development in order to achieve the desired results in terms of NMT penetration.

7.1.2.8 Policies to Conduct Awareness-Raising Campaigns

These are long-term commitments and hence require considerable and committed resources from the Government. Nongovernment organizations funded by international environmental agencies have a special role to play here. This would have good long-term benefits on environment and social criteria. Table 7.1 presents, indicatively, the above-discussed impacts of each policy measure considered.

Though each policy has its costs and benefits, for policy makers, in order to prioritize their choice, it is necessary to have a comprehensive idea on which policies stand better chances to be successful in achieving NMT implementation. Thus it is necessary to prioritize the policies. However, prioritization of policy measures would depend, highly, on the criteria based on which the comparisons are made. Thus, the policy measures are prioritized based on multiple criteria.

7.1.3 Methodology for the Prioritization of Policy Measures in Multi-criteria Approach

7.1.3.1 Establishing Evaluation Criteria

Criteria that represent various administrative, financial, behavioral, and societal factors should be chosen in order to have a comprehensive assessment of the alternative in a multi-criteria framework. Such representative criteria should be selected.

7.1.3.2 Multi-stakeholders

All actor groups involved in NMT and its planning and implantation should be included in opinion collection by means of a questionnaire toward the prioritization of policy measures. All such actors groups include Environmental Expert, policy maker, NGO, and Transportation Expert/Planner.

Table 7.1 Cost-benefits of policy measure to implement NMT

Policy and alternative measures	Costs on government	Burden on users	Employment generation	Environmental and health benefits	Congestion and other social benefits
Policies to integrate NMT with the public transport system planning at city/municipality level	√√√	–/√	√√√	√√	√√
Formulation of National Strategy for NMT as a facilitating framework for local plans	√	–	–	–	–
Policies to extend more specific incorporation of NMT in national transportation policy	√√√√	√	√√√	√√√	√√√
Incorporation of standards for NMT in new road infrastructure design	√	–	√	√√	√√
Policies to provide NMT-friendly infrastructure	√√√√	√	√√√√	√√√	√√√
Policies to provide soft loans and relaxing excise duty on importing bicycles/bicycle parts	√√	–	√	√	√
Policies to make regulations safeguarding NMT users and to cultivate respect for the NMT users	√	–	–/√	–/√	–/√
Awareness-raising campaigns	√√	–	–	√√	√√

"√√√√" indicates high significance and "–" indicates no impact

7.1.3.3 Scale

Performance of each policy alternative based on each criteria can be assessed on a 1–4 scale, with 1 representing "poor" and 4 representing "very good." A personal interview (questionnaire survey) method can be employed. The average of responses from all the respondents against each question can be used as the final

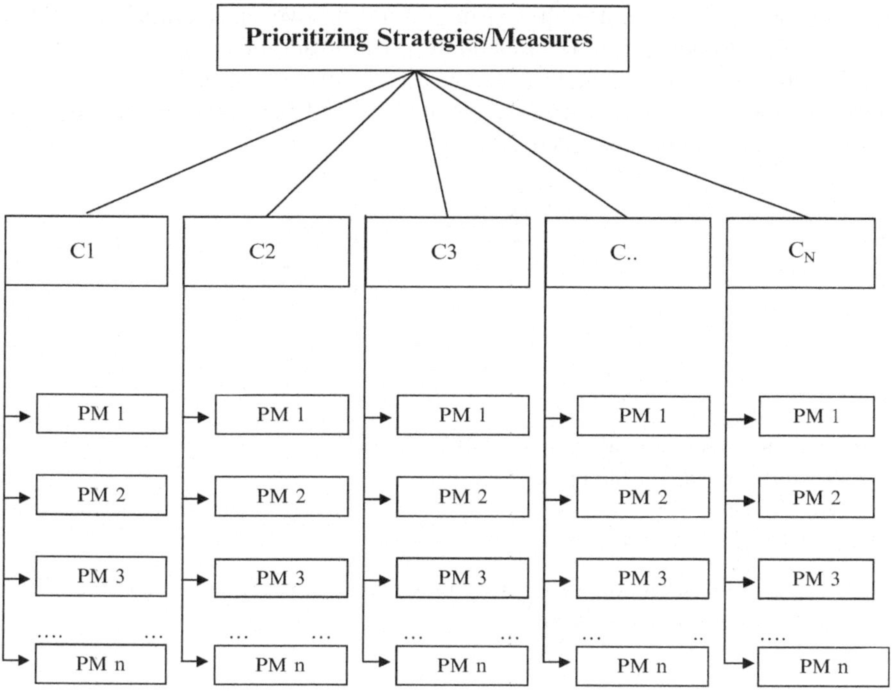

Fig. 7.1 Schematic presentation of prioritization of policy measures (PM) using multi-criteria approach

rating for the analysis. The policy measures can be finally assessed by aggregating their assessment based on individual criteria to arrive at final ranking. Figure 7.1 presents this generic process schematically. The same can be customized and developed for case-specific model.

The following scale can be used to rate the policy measures, qualitatively, based on different criteria. Intermediate values, viz., 1.5, 2.5, and 3.5, can also be used as and when required.

1	Poor
2	Moderate
3	Good
4	Very good

For instance, implementing a measure *policies to integrate NMT with the public transport system planning at city/municipality level* may need substantial improvement in the administrative capability; thus these policy measures may get poor score (say 1) based on the criteria "administrative capability." However, implementing the measure *policies to conduct awareness-raising campaigns* may not need any improvement in the existing administrative capacity; thus this PAM may get better score (3.5 or 4) based on the criteria "administrative capability."

Responses from all stakeholders can be used to arrive at common ratings for the policy measures under each criteria. Total score and percentage scores for each policy measure can be calculated as follows:

$$\mathrm{PM}_i = \left[\frac{\sum_{c_i=1}^{n} \left(r_{c_i} \times w_{c_i} \right)}{4} \right] \times 100$$

where

PM_i is the percentage total score of policy measure
r is the rating given for the policy measure
c_i is the criteria "i"
w_i is the weight on criteria "i"

The higher the score, the better the policy measure. PAMs with more than 50 % score can be considered positively.

7.1.4 Case Study

Based on the barriers for the implementation of NMT, as presented in Chap. 6, the following policy measures (PM) are identified for Indian metropolitan cities such as Mumbai and Delhi:

PM 1	Policies to integrate NMT with the public transport system planning at city/municipality level
PM 2	Formulation of National Strategy for NMT as a facilitating framework for local plans
PM 3	Policies to extend more specific incorporation of NMT in national transportation policy
PM 4	Incorporation of standards for NMT in new road infrastructure design
PM 5	Policies to provide NMT-friendly infrastructure
PM 6	Policies to provide soft loans and relaxing excise duty on importing bicycles/bicycle parts
PM 7	Policies to make regulations safeguarding NMT users and to cultivate respect for the NMT users
PM 8	Policies to conduct awareness-raising campaigns

The following *criteria* are relevant in prioritizing the policy measures:

Administrative costs

This includes physical infrastructure cost, cost of changing existing rules and regulations, capacity building of implementers and stakeholders, cost of hiring the consultants, cost of collecting required information, cost of program monitoring, etc.

Financeability

Financing barrier removal policies can be a major challenge for the developing countries. Hence, financeability was selected as another criterion to choose appropriate policy options.

Environmental and social benefits

Potential to reduce environmental emissions (both local and global), potential to streamline the traffic in general, employment generation, reduction in accidents, and more affordable transport for urban poor

Administrative capabilities

Policies should be practical and feasible to actually produce intended benefits. Implementation of barrier removal policies requires a good organizational setup with appropriate infrastructure, man power, and technical support.

Political willingness

This criteria expresses the willingness of political leaders to propose and promote NMT-friendly policies.

Multi-stakeholder responses are collected using a questionnaire from all actors involved, viz., Environmental Expert, policy maker, NGO, and Transportation Expert/Planner. The scale as presented in methodology is used to collect the options of stakeholders.

The methodology as explained is applied, and all the responses from stakeholders are used to arrive at common ratings for the policy measures under each criterion, which is presented in Table 7.2

Policies to conduct awareness-raising campaigns are found to be the most effective measures in order to implement NMT. *Policies to integrate NMT with the public transport system planning at city/municipality level* and *policies to incorporate standards for the bicyclist and pedestrian provisions in new road infrastructure design* followed the awareness-raising policies in their priority. It is interesting to observe that the three top-rated measures represent three categories of policies, viz., campaigning, short-term measures (infrastructure improvement), and long-term planning (incorporation of NMT facilities in the road design itself). It clearly highlights the need for an integrated approach in development activities.

7.1.5 Short-Term and Long-Term Measures for the Implementation of Non-motorized Modes of Transport

Based on the priority rating of policy measures in the case study considered, the following long-term and short-term measures may be considered for the promotion of NMT in Indian metropolitan cities.

Table 7.2 Rating of policy measures based on different criteria

Policy measure	Administrative costs	Financeability	Environment and social benefits	Administrative capability	Political willingness	Score
Weights	0.2	0.2	0.2	0.2	0.2	
Policies to integrate NMT with the public transport system planning at city/municipality level (**P1**)	2.5	3.5	3.0	2.0	2.0	2.6 (65 %)
Formulation of National Strategy for NMT as a facilitating framework for local plans (**P2**)	2.5	3.5	2.0	1.5	2.5	2.4 (60 %)
Policies to extend more specific incorporation of NMT in national transportation policy (**P3**)	2.0	2.5	3.0	2.5	2.0	2.4 (60 %)
Incorporation of standards for the bicyclist and pedestrian provisions in new road infrastructure design (**P4**)	3.0	2.0	3.5	2.0	2.5	2.6 (65 %)
Policies to provide NMT-friendly infrastructure (separate bicycle lanes, wider pedestrian ways with cycling facility, subways and overhead passages at major intersections, and bicycle parking places at all public transport stations and public places) (**P5**)	2.0	1.5	3.5	2.5	2.5	2.4 (60 %)
Policies to provide soft loans and relaxing excise duty on importing bicycles/bicycle parts (**P6**)	3.0	2.0	2.0	2.5	2.5	2.4 (60 %)
Policies to make regulations safeguarding NMT users and to cultivate respect for the NMT users among the law-enforcing officers (police) and the motor vehicle users (**P7**)	2.0	3.0	2.0	1.5	2.0	2.1 (52.5 %
Awareness-raising campaigns (**P8**)	2.5	2.5	3.0	3.0	3.0	2.8 (70 %)

NMT as an option in transportation works better when the access and egress legs are within the walkable and cycling distances. Such a system works in the presence of an extensive mass rapid transit system (MRTS) wherein the prime leg is catered by rail and/or road. Mumbai presents such a case that suits well for NMT. Delhi with its newly developed metro is yet to transform its system into a full-fledged MRTS in order to fall into the perfect NMT situation for its access and egress legs.

The prioritization exercise with a multi-stakeholder participation approach suggests that the policy measures, viz., policies to integrate NMT with the public transport system planning at local level, formulation of National Strategy for NMT as a facilitating framework for local plans, incorporation of standards for the bicyclist and pedestrian provisions in new road infrastructure design, policies to provide NMT-friendly infrastructure, policies to provide soft loans and relaxing excise duty on importing bicycles/bicycle parts, policies to make regulations safeguarding NMT users, and policies to conduct promotional campaigns, could be used for better adaptation of NMT in Indian metropolitan cities.

Based on multiple criteria of *administrative costs* (for the policy implementation), *financeability* (of the policy measure), *environmental and other social benefits*, *administrative capability* (required to implement the policy measure), and *political willingness*, policies to create awareness and capacity (building) get higher priority followed by policies to integrate NMT with the public transport system at city/municipality level and policies to incorporate standards for the bicyclist and pedestrian provisions in new road infrastructure design.

Through a careful examination of various policy alternatives considered for the case study and their ranking, characteristics of transport system in Indian cities, and social conditions, the following recommendations can be made for the better implementation of non-motorized transportation:

1. *Policies to provide separate bicycle lanes* (semi-separate to begin with but fully separate lanes for the roads planned for the future) and parking places at all metro (rail) and bus stations in the MRTS network. This is the basic policy requirement for NMT implementation in Mumbai.
2. *Promotional and awareness campaigns*: strong promotional/awareness campaigns and capacity-building programs need to be developed for successful implementation of NMT.

   Changing behavioral patterns is as difficult as providing exclusive infrastructure for NMT in short time. A fast-moving city like Mumbai cannot afford too much of a disturbance due to the infrastructure development works. Thus, for all practical purposes, intermodel integration provides a better solution. Humid climatic conditions and predominant presence of urban poor further support this observation.
3. *Intermodal integration*: Indian cities such as Mumbai present different levels of commuting, different levels of affordability, and different conditions under which people travel. It would be increasingly difficult to replace some modes with NMT. Instead it would be better if intermodal integration is attempted.

Various policy combinations that could be considered with least disturbance for the existing transport system are:

- Clear footpaths and signaled intersections/overpass crossings covering entire access leg
- Partly separated lane (painted) for bicycles with designated bicycle parking without fee, next to the rail/bus stations
- Partly separated lane (painted) for bicycles with designated parking places with charges integrated in the monthly bus/train cards/passes
- Bicycle rentals adjacent to the rail/bus stations with parking fee integrated into the monthly train/bus pass and partly separated (painted) bicycle lanes
- Increase in the initial fare of auto-rickshaws and providing of indirect access to transit points with shorter routes earmarked for bicycles with a parking space without fee next to the rail station

As these modals need less of physical infrastructure and more of institutional and financial arrangements/mechanisms, it presents a high likeliness of success. These measures could be *short but immediate measures to kick off with NMT intensification*. Higher affordability in metropolitan cities can be effectively captured with bicycle storage facilities with parking fee.

As long-term measures, according to the findings of the case study, NMT needs to be integrated into the existing public transportation system of the city. One potential way of doing that is to make the access leg completely catered by NMT. This essentially indicates the development of "bike and ride" modal for access leg of daily commuting.

4. *Developing "bike and ride (and bike)" model commuting*: with the developing country cities such as Indian cities constructing their metro systems, it would be better to complement it with a well-designed NMT integrated transit system. With an average access leg followed by primary leg and much shorter egress leg, policies to replace the exciting "ride and ride (and ride)" (in line with the "park and ride" modal) mode commuting by "bike and ride (and bike)" mode would lead to successful adaptation of NMT.
5. *Bicycle lending system and the necessary financial mechanisms*: though providing bicycle lanes and parking at mass transport transit points help people use bicycle for the access leg, it is essential to cater for the egress leg also to make it a complete loop of "bike and ride (and bike)." Thus, bicycle lending system needs to be introduced at important business centers. This needs to be supported by various financial mechanisms and incentive structures. This along with motor vehicle restrictive policies in the business district[1] would result in faster adaptation of NMT. The major social benefit of this system is huge employment generation, particularly for the urban poor. Thus, it adds to one of the major Millennium Development Goals – poverty eradication in urban centers.

[1] Like car-free Sundays, restricted entry for motor bikes in the central business district, allowing odd and even number plated vehicles on alternate dates, etc.

This system, for its successful implementation over a longer period of time, needs to be complemented by the policies to internalize the NMT provisions and standards in transportation planning and design.

6. *Need for cost-benefit estimation of intermodal integration*: intermodel integrated policies need to be assessed for their benefits in terms of affordability of transport for low-income urban groups (as % income spent), reduction in congestion, energy saving, and control of CO_2 and other emissions. In the case of multi-model integrated access, fee structure needs to be fixed based on the affordability study and a complete cost-benefit analysis of the "policy model." In this context, Global Environmental Facility (GEF) projects can be encouraged as this would involve substantial GHG reductions. The following section presents such an attempt at finding co-benefits of the system.

7.1.6 Costs and Benefits of Initiative to Promote NMT in Mumbai: An Indicative Presentation

This specific case study of co-benefits in transport system is conducted for transportation system of Greater Mumbai City with very specific NMT compliance initiatives. However, this may generally be used as a guideline for such benefits from the other cities as well.

While it is important to identify and prioritize specific policy alternatives and measures to promote NMT and their implementation, it is also important to build the basic capacity and infrastructure as a first step toward achieving such idea of NMT in Mumbai. Two such requirements are:

- Sensitization of top officials and providing basic and advanced training for different levels of man power involved in urban transportation and allied departments.
- Developing basic infrastructure, viz., roads with proper footpath and dedicated bicycle lane (could be joined with footpath in some cases), providing bicycle parking places at all metro train stations (73 × 2), and improving road intersections making them conducive for NMT users.

Providing these two basic blocks would result in various costs and benefits (and some co-benefits as well). It was estimated that capacity building in Mumbai by means of sensitization of top officials and training for middle- and lower-end officials would cost about 178,382 USD. This kind of training program may be called for in every 5-year interval, after making a systematic need assessment study.

Provision of basic infrastructure includes widening of the existing network of urban roads (1,973 km) and providing proper footpath and NMT lanes, improving the existing road intersections to support NMT users, and developing bicycle stands at all 73 metro rail stations (one bicycle stand on east and west sides of each station). Cost of providing footpath and NMT lane on Mumbai roads is calculated.

Complete conversion of the existing road network to provide NMT in Mumbai is expected to cost about 51 million USD.

A 10 % conversion of roads to NMT-friendly roads in Mumbai (with bicycle stands at all metro rail stations and improved road intersections) would cost 15 million USD whereas 100 % conversion would cost as much as 136 million USD.

Promoting NMT has many benefits, viz., reduced air pollution, reduction in CO_2 (greenhouse gas) emissions, energy saving, slowdown in motorization, reduction in congestion, and other social benefits as explained in the main text. To provide an indicative idea of such benefits, reduction of CO_2 and other environmental emissions was calculated in the present study.

Conversion of 10 % roads into NMT-friendly infrastructure (which is expected to result in 10 % reduction in the number of three-wheelers over the next 5 years) is estimated to result in a reduction of CO_2 emission by 7.63 million tons over the next 5 years. NMT promotional efforts with a possible 75 % reduction in the number of three-wheelers over the next 5 years would result in reduction of 19.8 million tons of CO_2 over the same time period.

At 10 % conversion of roads to NMT-friendly roads, the marginal abatement cost (MAC) of carbon dioxide was found to be as low as 1.97 USD/ton of carbon. At 100 % conversion of roads (75 % reduction in three wheelers stock) to NMT-friendly roads, MAC was found to be 6.87 USD per ton of CO_2 reduction, making it a likely initiative for future CDM activity. NMT is expected to result in substantial co-benefits in terms of reductions on carbon monoxide, hydrocarbon, and oxides of nitrogen emissions.

An indicative estimation of cost benefits of initiatives to promote NMT viz. capacity building and provision of infrastructure revealed that promotion of NMT is has substantial benefits both in the form of GHG and local emissions control. Marginal Abatement cost was found to be in the range of 2–7 USD per ton of carbon reduced. Initiative of providing basic infrastructure like converting the existing roads into NMT friendly ones, bicycle stands at all rail stations and modernizing road intersections for NMT usage was found to be costing in the range of 15–136 million USD (Yedla 2007).

7.2 Policy Measures to Promote Cleaner and Alternative Modes of Transportation

Policy analysis considers not only the impacts of the policy measures on the prime objectives but also the ancillary benefits as presented in the previous section. In a similar approach to prioritize policy measures, a modified methodology is proposed which attributes more weight for the policies that can address more than one barrier. The weight is a function of the number of barriers that can be addressed by the policy measure.

7.2.1 Methodology for the Prioritization of Policy Measures Based on Multi-criteria Framework

Four necessary steps for any policy analysis are:

- Defining the problem
- Identification of the policy alternatives
- Establishing evaluation criteria
- Evaluating alternative policies

7.2.1.1 Defining the Problem

The objective of the analysis is to suggest potential policy measures to overcome major barriers to the adoption of alternative technological options that were discussed and analyzed in Chap. 4.

7.2.1.2 Identification of Suitable Policy Measures

There are a variety of policies discussed in literatures for removing barriers to widespread adoption of cleaner options. A policy matrix needs to be developed with a set of common policy measures, which have potential to remove one or more barriers to the selected alternative options. As some strategies could be common across barriers, it is possible to arrive at common policy measures potential in removing one or more than one barrier. Policy measures have to be classified according to their potential in handling more than one barrier and assign weights accordingly. Heavier weight should be given to common policy that can be applied toward removal of more than one barrier.

The selection of policy measures could be based on review of various existing policies and successful demonstrations in foreign countries. Personal interviews/brainstorming sessions among stakeholders involved could reinforce the primary selection based on review. Weights can be derived based on normalization of their potential in handling a number of barriers.

7.2.1.3 Establishing Evaluation Criteria

In order to compare the merits and demerits of various policy alternatives, a set of criteria needs to be identified. Relevant evaluation criteria can be selected based on the candidate policy measures as most of the criteria are policy dependent.

7.2.1.4 Weighing Alternative Policy Measures

It is necessary to assess the benefits each policy alternative could contribute corresponding to the identified criteria. Detailed assessment of costs and benefits of these policy alternatives is resource and time intensive. Given the resource and time

Table 7.3 Sample structure of a criteria/policy alternative matrix

Criteria	Score	Policy alternatives Alternative 1	Alternative 2	Alternative 3	Alternative 4
Criteria 1	0.3	2.5	3.0	2.0	3.5
Criteria 2	0.25	3.0	2.0	3.5	2.5
Criteria 3	0.25	3.5	3.0	3.5	2.0
Criteria n	0.2	3.0	3.5	2.5	2.0
Total weighted average score	1	2.975	2.85	2.475	2.575
Percentage total weighted average score		74.4	71.25	61.9	64.4

constraints, in many cases, it may not be possible to quantify or fully evaluate policy costs. Costs may also depend on specific policy options promoted and the means of implementation. Evaluation of the cost of policies is complex and has not been documented by developing countries and is not available in the literature. In the absence of these, the cost valuation might not be completed. With this limitation, one may give indicative values of policy costs during the evaluation process. A descriptive approach can be adopted in the evaluation process.

This valuation can be used to assign weights to the policies that are already selected based on the "review" and the input of experts. Hence, this could even be qualitative and descriptive. Hence, the quantification of impacts, which is in general time and resource intensive, is minimized. The main task in evaluating policies is filling in all the cells of the matrix, as shown in the demonstration table (Table 7.3) with values and estimates and projected impacts of the policy under each criteria. These predictions may sometimes be quantitative especially when available information allows us to measure impacts in terms of numbers. Sometimes the impacts predicted shall be qualitative, i.e., "poor," "good," "very good," or "excellent."

The matrix is constructed based on the opinions and judgments from the group of experts in the field by means of brainstorming sessions/personal interviews. Results of the interviews conducted are translated into scales (e.g., poor = 1, good = 2, very good = 3, and excellent = 4). The average response will be the final value that shall be reflected in the final matrix. The final matrix gives a picture on how each alternative performs relative to the pre-identified criteria. The resulting preference for the alternative depends on weight distribution of the criteria as described above. Weights for the criteria can be obtained in a similar method. The total weighted average score of each alternative is calculated (Yedla 2007):

$$\mathrm{PM}_i = \left[\frac{\sum_{c_i=1}^{n} (r_{c_i} \times w_{c_i})}{4}\right] \times 100$$

where

PM_i is the percentage total score of policy measure.
r is the rating given for the policy measure.
c_i is the criteria "i".
w_i is the weight on criteria "i".

As different policy alternatives have different capacities to handle a number of barriers, these scores are further subjected to the normalized weights derived for the policy alternatives based on their capacity to handle the number of barriers. This final weighted score would reveal the actual potential of these alternative policies. The policy alternative with highest score may be selected as the most appropriate policy among the alternatives.

7.2.2 *Case Study*

This case study considered Mumbai and Delhi which could be representative case for the rest of the cities in India. However, the analysis could be treated generic and can be used as an indicative presentation for all metropolitan cities.

Most common barriers for the implementation of cleaner fuel and other alternative options in transport sector are presented in Chap. 6. The same list has been further augmented by means of thorough literature review, and consultation with stakeholders is given below in order to identify suitable policy measures to control them:

- Lack of resources and infrastructure (**B1**)
- Additional cost (**B2**)
- Lack of enforcing mechanism (**B3**)
- Lack of awareness (**B4**)
- Lack of availability of efficient technology/conversion kits (**B5**)
- Unwillingness to participate (**B6**)
- Lack of proper training and implementation programs for smooth transition (**B7**)

Based on the above set of barriers and the review of policies in the literature also presented in Chap. 8, a set of policy alternatives are considered against each barrier and presented in the table below (Table 7.4).

7.2.2.1 Selection of Alternative Policies and Their Weights

Among the most important barriers, *lack of resources and infrastructure* is ranked as the most prominent barrier to the adoption of cleaner fuels in India, with particular emphasis on natural gas vehicles. Infrastructure for the supply of cleaner fuels is essentially filling stations and pipe network in the case of CNG and desulfurization units in the refineries for the supply of low-sulfur diesel and

Table 7.4 Policy alternatives for each barrier to the adoption of cleaner alternatives in Delhi and Mumbai

Barrier	Policy measures (PMs)
Lack of resources and infrastructure (**B1**)	Formulation of an implementing agency to provide the necessary infrastructure ($\mathbf{P_{11}}$)
	Partnership between the government, public sector undertakings, and private actors in providing the infrastructure ($\mathbf{P_{12}}$)
Additional cost (**B2**)	Subsidizing the additional expenditure ($\mathbf{P_{21}}$)
	Interest subsidies on loan to convert vehicles ($\mathbf{P_{22}}$)
	Financial incentives such as free or priority parking and separate lanes for alternative fuel vehicles (AFVs) ($\mathbf{P_{23}}$)
	Free inspection and maintenance ($\mathbf{P_{24}}$)
	Access to the otherwise restricted areas ($\mathbf{P_{25}}$)
	Tax on polluting vehicles and earmarking such revenue to provide subsidies for AFVs ($\mathbf{P_{26}}$)
Lack of enforcing mechanism (**B3**)	Single window/priority checkpoints ($\mathbf{P_{31}}$)
	Task force to carry out checks ($\mathbf{P_{32}}$)
	Heavier fines on defaulters ($\mathbf{P_{33}}$)
	Coloring scheme for easy tracking of vehicles ($\mathbf{P_{34}}$)
Lack of awareness (**B4**)	Awareness campaigns ($\mathbf{P_{41}}$)
	Demonstration of AFVs and their advantages ($\mathbf{P_{42}}$)
Lack of availability of efficient technology/conversion kits (**B5**)	Development of indigenous technologies ($\mathbf{P_{51}}$)
	Creation of market by command and control measures ($\mathbf{P_{52}}$)
	Time-bound waivers on import duty ($\mathbf{P_{53}}$)
Unwillingness to participate (**B6**)	Command and control measures ($\mathbf{P_{61}}$)
	Judicial interventions ($\mathbf{P_{62}}$)
	Demonstration of AFVs and their advantages ($\mathbf{P_{63}}$)
Lack of proper training and implementation program in case of CNG technology (**B7**)	Training programs for the workers ($\mathbf{P_{71}}$)
	Awareness campaigns for the drivers ($\mathbf{P_{72}}$)
	Demonstration of AFVs and their advantages ($\mathbf{P_{73}}$)

Note: Number in parenthesis refers to the barrier it can address followed by the serial number

network of charging stations for battery-operated vehicles. To remove this barrier, the efforts from the Government need to be complemented by the market. Thus, it is proposed that there be a partnership between the City Government, public sector undertakings, and the private actors. This kind of integrated body would have the authority of the government and also participate in the market.

There is a need for a dedicated and exclusive body to provide the infrastructure and services so as to remove the lack of infrastructure for the penetration of cleaner alternatives. Indraprastha Gas Limited is one such example in Delhi. Such bodies mobilize their own resources from various stakeholders, and by going in for public issue, they can sustain the burden and also clear the financial burden in a relatively shorter span of time. This barrier can be handled effectively by the above two measures. Just by providing subsidy to the service provider, it may not be possible to create the momentum for its spread and longevity.

Additional cost is another most important barrier to the adoption of AFVs. Conversion of existing vehicles to CNG needs substantial investments. The cost of a diesel-CNG kit is about Rs. 85,000 (Indian rupees) and that of a petrol-CNG kit is about Rs. 30,000, exclusive of customs duty. This investment requirement has become a major barrier to its adoption. A new CNG bus costs around 1.6 million Indian rupees whereas the diesel version costs only 0.8 million Indian rupees. These additional expenditures overshadow the fact that the CNG is cheaper and is available at a price in gasoline-equivalent terms of around Rs. 16.83/liter, which compares favorably with gasoline prices of Rs. 31.49/liter and diesel prices of Rs. 20.12 in April 2003. It is estimated that the costs of converting a vehicle to CNG can be recovered after 30,000 km. However, the feeder services like three-wheelers (auto-rickshaw) face severe problem in meeting this huge initial cost of conversion kits. Hence, it is essential to provide some sort of financial incentives to overcome this barrier and allow the user to capture the low operation cost from the AFVs.

In case of major manufacturers, it is possible to capture the market demand, and that essentially is the driving force. Demand-driven markets take care of the additional investments required. In India, the CNG market is limited to a total of around 7,000 vehicles in Mumbai and Delhi, with total demand of around 0.03 MMcm/day in 2001. This number has gone up substantially and reached as high as 70,249 motor vehicles in Delhi alone creating an active market for natural gas vehicles (NGVs). However, the small-scale user needs incentives to overcome this important barrier.

A subsidy in the range of 50 % on the additional cost for AFVs would encourage conversions to a great extent. However, it puts a heavy burden on the Government. An alternative measure could be interest subsidies on loans to convert the existing vehicles to CNG or other alternative fuels. Some levels of command and control measures are required to break the stigma and to gain the momentum, and these financial incentives would add to that momentum. In such an approach in Delhi to increase the penetration of CNG vehicles, the Government of India (GoI) has provided an exemption of customs duty on CNG converter kits imported into the country. Incentives attempted include an exemption in the 12 % state sales tax and a 3 % low-cost loan option for the weaker sections of the society. Customs duty on imported CNG kits was waived. This has resulted in a decline in the price of a converter kit by about 15 %. However, these subsidies were given for initial small period, and they were removed subsequently. Interest subsidy was claimed only by 60 cases in Delhi, which is highly insignificant. Similar mechanisms can be tried to improve the penetration of CNG in Mumbai and other cleaner options like electric and battery-operated vehicles both in Delhi and Mumbai.

Unlike in the case of Delhi, Mumbai misses the advantage of having regulation for the conversions, though Mumbai high court has issued directives. Hence, there is a need for indirect incentives like free or priority parking for vehicles on alternative fuels and providing them with special lanes, which essentially results in speedy travel and fuel savings. These mechanisms have been attempted successfully in many developed countries.

Other indirect benefits that can be offered to minimize the friction due to additional costs are free inspection and maintenance and access to areas like CBD where other gasoline and diesel vehicles are prohibited. Providing with free inspection and maintenance would facilitate with ease in monitoring the conversion of vehicles, and that makes the enforcement that much more effective. Cleaner and alternative fueled vehicles can be prioritized to enter certain restricted zones in the city as they are nonpolluting. This would be a major driver, in spite of additional costs involved for switching to cleaner fuels.

Another more direct and effective financial incentive could be a tax on polluting vehicles and earmarking such revenue to provide subsidies for AFVs. This would give a positive impression on the person affected. This essentially neutralizes the pollution and adoption of cleaner fuels. This can be affectively applied in the case of Mumbai where the defaulters can be easily captured due to the geographical features of the city. This measure leaves behind negligible burden on the state.

Among various barriers to the adoption of AFVs, *lack of enforcing mechanism* is important because this directly influences the enforcement of the alternative. General resistance to any change causes the most potential alternative also to fail. In spite of providing infrastructure and financial incentives, the alternative options may fail in the absence of enforcing mechanisms.

Once the regulation for conversion of vehicles is made, a single window/priority checkpoint would enhance the effectiveness of enforcement. A task force to carry out random checks on vehicles would remove this barrier of weak enforcement. As an indirect approach, levying heavier fines on defaulter could be an effective enforcement tool. This would work well in the case of three-wheelers and cars. For buses the enforcement will not be a barrier as the number is limited and easy to monitor.

A scheme of distinct coloring for AFVs would help in easy tracking of vehicles. This would enhance the enforcement, and fining defaulters would be that much easier by employing this scheme.

In the case of semipublic transport encompassing taxis and three-wheelers (autorickshaws), *lack of awareness* also poses a major barrier. Vehicles driven on natural gas have almost half the operating cost of the conventional fueled vehicles. Initial capital cost requirements overshadow other benefits because of lack of awareness.

Hence, removing this barrier could help AFVs penetrate well. Owing to the stage of advancement, people are not aware of battery-operated vehicles and their benefits. Hence, it is necessary to design and execute strong awareness campaigns like billboards, advertisements in newspapers, and other media like TV and radio. NGOs have a particularly important role in this. Another measure to counter this barrier could be demonstration runs of these alternative fueled vehicles in the city and their advantages.

In the initial phase of the AFVs, *lack of availability of efficient technology/ conversion kits* poses a severe constraint on their penetration. However, this barrier disappears over a period of time as the market itself takes care of such demand. However, it is essential to reduce its intensity to achieve a better penetration rate in the beginning. This barrier can be controlled in a three-pronged approach. During the initial period, incentives like waiver on import duty on conversion kits should be given to encourage the conversion of vehicles. However, it needs to be a time-bound process slowly reducing the waivers so that it drives the development of indigenous supply of kits. However, leaving it to the natural market might make it unsuccessful at times as it takes long to create a market. Hence, it is necessary to make a regulation of this conversion so that the market is created with a specific time frame. Therefore, there is a need for making regulation to make the conversions mandatory. Subsequently, to meet the demand, indigenous technology and conversion kits need to be developed over a period of time. These three measures put together can remove this barrier of nonavailability of efficient technology/ conversion kits.

One of the most difficult barriers to handle in case of alternate fuel vehicles (AFVs) is people's *unwillingness to participate*. There is always a resistance to the adoption of new methods when people got used to the certain mode (s). Most of the time market fails to remove this barrier. In spite of financial incentives and other measures, people resist a change. There are only two possible ways to handle this problem both in Delhi and Mumbai, and they are command and control with judicial intervention and demonstration of AFVs and their advantages to individuals and public in general.

As in the case of Delhi, a strict regulation to adopt AFVs is necessary to force the conversions in spite of people's unwillingness to participate. This is essential especially when the environmental situation is dire in the city and needs to be attended. Vehicle conversion to clean fuels needs to be made mandatory with a strict time frame. Demonstration runs of these alternate fuel vehicles (AFVs) and their benefits would reduce the initial friction resulting from the regulation. Hence, it is essential to organize such demonstration runs of AFVs and their advantages to the individual owners and also the co-benefits like improved air quality to the general public.

Finally, lack of proper training and implementation programs for smooth transition is a functional barrier affecting the efficiency of conversion/penetration. As such alternatives as natural gas and battery-operated vehicles are technology intensive, the workers handling them need to be trained specially for the task.

Table 7.5 Matrix of barriers and PMs

Barriers	Policy erasures
Lack of resources and infrastructure (B1)	Formulation of an implementing agency to provide the necessary infrastructure
Additional cost (B2)	Partnership between the government, public sector undertakings, and private actors in providing the infrastructure
Lack of enforcing mechanism (B3)	Subsidies on the additional expenditure
Lack of awareness (B4)	Financial incentives like free or priority parking, separate lanes for alternative fuel vehicles, and free inspection and maintenance
Lack of availability of efficient technology/conversion kits (B5)	Tax on polluting vehicles and earmarking such revenue to provide subsidies for AFVs
Unwillingness to participate (B6)	Single window/priority checkpoints
Lack of proper training and implementation programs for smooth transition (B7)	Task force to carry out checks
	Heavier fines on defaulters
	Coloring scheme for easy tracking of vehicles
	Awareness campaigns
	Demonstration of AFVs and their advantages
	Time-bound waivers on import duty
	Regulations
	Training programs for the workers
	Awareness campaigns for the drivers

Lack of such training would fail to provide a smooth transition to the AFVs. Hence, this needs to be attended to make the AFVs' penetration more effective and fast.

Delhi has experienced this barrier in the case of natural gas vehicles. No proper training was given to the workers involved in handling the AFVs, and that has resulted in various operational problems. Improper conversions lead to mishaps creating havoc in the public. This barrier is very prominent during the initial phase of the induction. Hence, it is essential to execute training programs for the workers involved.

Every new technology involves some adjustments to the existing system, and drivers should be made to understand that in the absence of which there is every possibility of mishandling of the vehicle. Hence, awareness campaigns for drivers (for taxis and auto-rickshaws in particular) are necessary to provide a smooth transition to these AFVs. Demonstration of AFVs and their advantages would reduce the gap between introduction and implementation of these efficient alternatives in urban transportation.

Certain policy alternatives and measures are common to few barriers. These measures have potential to handle more than one barrier. Any such measure handling more than one barrier should be given priority over the other which handles only one barrier. Thus, a pool of policy alternative and measures are selected against the set of barriers. Policy measures listed against each barrier in Table 7.4 are further clustered to make a final set of policy measures. Table 7.5 presents this matrix of barriers and PAMs.

Table 7.6 Scores for policy measures

PMs	Affected barriers	Normalized score
Formulation of an implementing agency to provide the necessary infrastructure (P1)	B1, B5, B7	0.75
Partnership between the government, public sector undertakings, and private actors in providing the infrastructure (P2)	B1, B2, B5, B7	1
Subsidies on the additional expenditure (P3)	B2, B3	0.5
Financial incentives like free or priority parking, separate lanes for alternative fuel vehicles, and free inspection and maintenance (P4)	B2, B3, B4	0.75
Tax on polluting vehicles and earmarking such revenue to provide subsidies for AFVs (P5)	B2, B3, B5	0.75
Single window/priority checkpoints (P6)	B3, B7	0.5
Task force to carry out checks (P7)	B3, B4, B5	0.75
Heavier fines on defaulters (P8)	B3, B2, B5	0.75
Coloring scheme for easy tracking of vehicles (P9)	B3, B4	0.5
Awareness campaigns (P10)	B4, B2	0.5
Demonstration of AFVs and their advantages (P11)	B4, B6, B7	0.75
Time-bound waivers on import duty (P12)	B5	0.25
Regulation (P13)	B2, B3, B6	0.75
Training programs for the workers (P14)	B7	0.25
Awareness campaigns for the drivers (P15)	B7, B3, B2	0.75

In an attempt to provide weights to different PAMs based on their potential to handle the number of barriers, various barriers are listed against each PAMs and weights are given by normalizing (maximum point normalization) their capacity to handle the number of barriers. These weights would be applied later in the process of deriving final performance of these PAMs in controlling barriers. Table 7.6 presents the normalized scores of each PAM. Consultations with experts (from the list given at the end) helped in arriving at these weights.

7.2.2.2 Criteria for Policy Analysis

A set of criteria is required to make a comparison between various policy alternatives and measures so as to assess their potential in removing as many barriers for better penetration of alternate fuel vehicles. In this section, important criteria are discussed and selected for further analysis.

Economic Considerations

Direct cost of a barrier removal policy may include administrative cost, cost of financial incentives, and cost of advertising the program. Impact of these policy measures on the transport sector and economy in general should be identified to assess indirect policy costs and benefits. However, the macroeconomic assessment is often time-consuming and requires extensive data. Because of the time and resource limitation in the present project, it is agreed to examine some factors, which have significant impact, by giving indicative values using descriptive approach.

Successful implementation of barrier removal policies include the following cost items:

- Administrative costs – this includes physical infrastructure cost, cost of changing existing rules and regulations, capacity building of implementers and stakeholders, cost of hiring the consultants, cost of collecting required information, cost of program monitoring, etc.
- Cost of financial incentives – financial incentives are one of the most commonly used barrier removal policies used to increase the wider adoption of CEETs. If the policies and measures include any financial incentive program, the cost of this incentive will have to be considered in addition to other administrative cost items.

The following are a few more important criteria that could play an important role in evaluating alternative policy measures in the cities selected:

Financeability: financing barrier removal policies can be a major challenge for developing countries. Clearly, financeability of the program could also be another criterion to choose appropriate policy options.

Administrative feasibility: policies should be practical and feasible actually to produce intended benefits. Implementation of barrier removal policies requires a good organizational setup with appropriate infrastructure, man power, and technical support. Therefore, administrative feasibility could be considered as one of the evaluation criteria.

Political acceptability: it may be difficult to get political support for most of the GHG reduction policies and measures because the policy makers are more likely to have their priority on economic and social needs, as mentioned earlier, rather than giving attention to environmental issues of global concern. The passing of GHG mitigation policies through political and bureaucratic process can be a challenge for developing countries. Hence, political acceptability could be one of the evaluation criteria.

Equity: equity can be considered as an evaluation criterion because the equitable distribution of policy costs and benefits among stakeholders could be of great importance.

Method of subjective comparison is applied in choosing the appropriate criteria to evaluate policy alternatives. A weightage matrix for the criteria was constructed based on the brainstorming session held at IGIDR and also consultations with experts. A group construction approach with a supra user was adopted in constructing this matrix.

Table 7.7 Derived weights for criteria

Criteria	Weight
Administrative cost	0.1
Financial burden	0.2
Human resource benefits	0.1
Administrative capability	0.3
Political acceptability	0.3

7.2.2.3 Selection of Criteria

Based on careful examination of the set of policy alternatives and measures (PAMs) considered, the following criteria are selected for further assessment:

- Administrative cost
- Financial burden
- Human resource benefits
- Administrative backup
- Political acceptability

Financial burden is the cost involved in meeting costs arising from the policy to remove barriers by means of financial incentives. Human resource benefits include employment generation, training of workers, development of skilled labor, etc. Following the group construction with *a supra user approach*, the following matrix of weights was constructed for the above-listed criteria (Table 7.7).

7.2.2.4 Prioritization of Policies Based on Criteria

All the policy measures (PMs) are analyzed based on the above set of important criteria. Any alternative option has strengths and weaknesses, and it is important to make a consolidated assessment accounting for all its strengths and weaknesses. Hence, the effectiveness of each PM against each criteria is assessed on a four-point scale as shown below:

Low, 1; medium, 2; high, 3; very high, 4

On this scale low represents the least impact of a policy in removing the selected list of barriers. To have the uniform representation and deriving scores, qualitative assessment was adopted for all PMs. However, quantitative analysis was made the basis for their representation on the scale presented for those cases where data is available. A group evaluation matrix was developed based on brainstorming among experts, and a supra user approach is applied while generating the matrix. Inputs were taken from various important stakeholders in urban transport sector to make the assessment complete.

All PMs are explained under each criteria and presented below.

Administrative Cost

Any policy to be implemented essentially involves various efforts like changing existing rules and regulations and organizational setup. If the policy cannot be implemented with the existing organizational setup, it takes quite an effort to make it work. It involves costs like infrastructure, hiring consultants and staff, training and education of the staff, etc.

Formulation of an implementing agency to provide necessary infrastructure essentially involves huge administrative costs. Construction of 23.8 km CNG pipeline in Delhi by IGL needed 250 million Indian rupees. Establishment of 110 CNG stations and the other necessary infrastructure needed 5,220 million rupees. It essentially involves other huge costs like training personnel and providing support in execution. Policies to promote partnership between major stakeholders also demand huge administrative costs. At times it would be a necessity to go for amendment of rules to accommodate such new bodies.

However, providing subsidies and financial incentives would not involve much administrative cost as these can be implemented with the existing organizational setup. Taxing pollution vehicles needs legislation in support, and such changes demand huge administrative costs. Similar costs would be observed in the case of fines on defaulters. Priority checks with single window and constitution of task force do not demand too much of administrative costs as they can be handled with the existing organizational setup and man power.

Color coding schemes, training, and awareness programs can be successfully implemented with the existing system without having many changes. Delhi presents an interesting illustration of achieving such without any significant additional costs. Time-bound waiver on import duty and regulation to promote AFVs involves substantial administrative costs.

The rating of policies under this criterion is presented below which is also presented in Fig. 7.2.

Financial Burden

Financing the partnership among key stakeholders is not difficult because of the fact that market takes care of the per capita demand. Financeability should not be a problem with such partnerships in place. Though formulation of implementation body involves substantial administrative costs, its financeability should be taken care by the shares held by individual actor.

Subsidies on the conversion present a serious case of financeability problem. Price difference between diesel driven bus and CNG bus is 0.8 million rupees with diesel bus costing 0.8 million rupees and CNG bus 1.6 million rupees. The cost of diesel-CNG conversion kit is Rs. 85,000 and petrol-CNG conversion kit, Rs. 30,000. The following table (Table 7.8) explains the quantum of burden that the state has to bear at various levels of subsidy given.

Criteria	P1	P2	P3	P4	P5	P6	P7	P8
Administrative Costs	1.0	2.0	3.0	3.0	2.0	3.0	3.0	2.0
	P9	**P10**	**P11**	**P12**	**P13**	**P14**	**P15**	
	4.0	2.0	2.0	2.0	2.0	3.0	3.0	

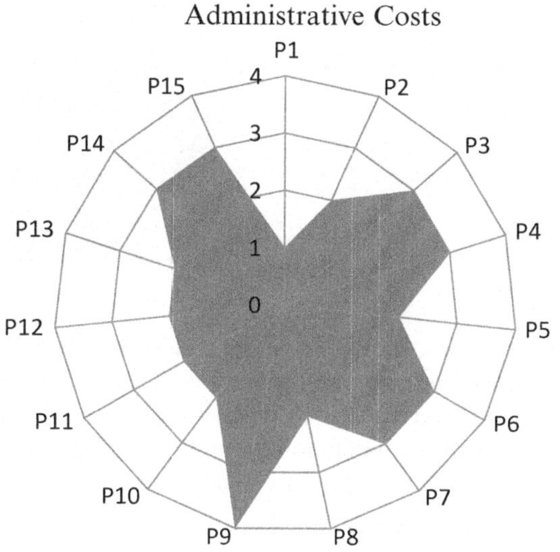

Fig. 7.2 Performance of policy measures on the criteria of administrative cost

Table 7.8 Quantum of burden that the state has to bear at various levels of subsidy given

Vehicle type	Number	100 % subsidy on conversion	50 % subsidy on conversion	25 % subsidy on conversion
Old bus conversion	7,231	614.6 million rupees	307.3 million rupees	153.6 million rupees
New CNG buses	2,539	2,031.2 million rupees	1,015.6 million rupees	507.8 million rupees
Three-wheelers	50,830	1,524.9 million rupees	762.45 million rupees	381.22 million rupees
Mini buses	4,468	134.04 million rupees	67.02 million rupees	33.51 million rupees

Subsidy essentially creates more penetration of the option. However, it places a heavy burden on the state as it runs into billions of rupees. Hence, this policy of providing heavy subsidy on conversions may not be viable as compared to other policy measures in spite of its potential in handling the barrier of "additional cost." Providing financial incentives also would lead to similar consequences and hence is rated low on this criterion.

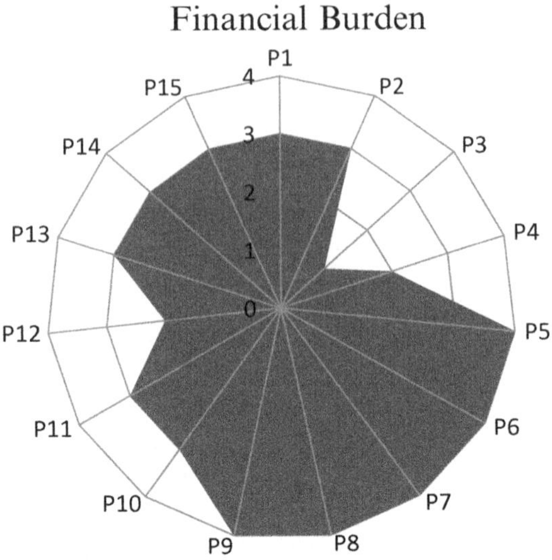

Fig. 7.3 Performance of policy measures on the criteria of financial burden

Tax on polluting vehicles, providing single window checkpoint facility, constituting task force for checking and implementation, heavier fines on defaulters, and distinct color scheme should not pose considerable financial strain on the state exchequer. In Delhi CNG implementation for public transport, many such activities are handled with the existing man power and organization without incurring significant resources. In the year 2002–2003, the daily fines on defaulter in Delhi amounted to Rs. 24 million.

Awareness campaigns and demonstrations need to be financed by the state, and its requirements depend on the level of activity. In a city where the pollution levels are extremely high, intensive campaigns required would entail considerable resources and hence financeability could be a problem. Time-bound waivers would face financeability problem as they involve huge loss in income for the state. Figure 7.3 presents the ratings assigned to various policies under this criterion of the financial burden on the state.

Human Resource Benefits

Certain policy measures such as establishment of implementing agency and partnership between the government, public sector undertakings, and private actors have very high potential to generate employment at various levels starting from labor technicians to managerial executives. However, partnership among various stakeholders such as government-owned agencies responsible for public transport, publicly owned transport corporations and the company(s) responsible for providing clean fuel, implementing body(s), and the local government has more potential in employment generation.

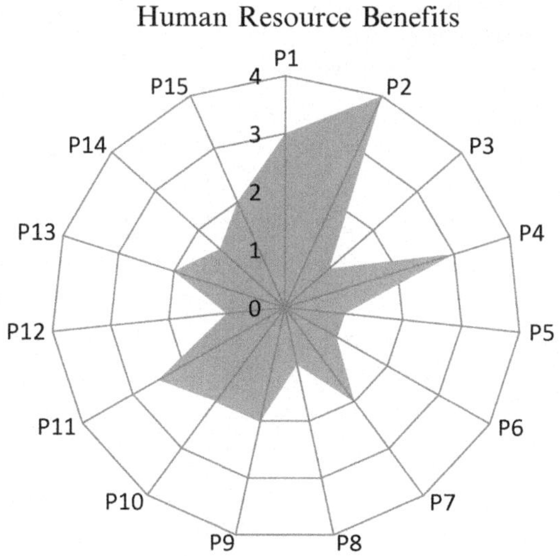

Fig. 7.4 Performance of policy measures on the criteria of human resource benefits

Policy measures such as free inspection and maintenance will have a moderate level of employment generation, which at times may be insignificant. In the case of Delhi CNG program, the additional responsibility of increased checks was managed by the existing staff of the transport department of Delhi. However, a city with a heavy vehicular population essentially shows potential employment generation. In Delhi CNG program 2.4 million vehicles were checked for pollution during the year 2002–2003. Taxing the polluting vehicles and single window scheme for pollution checks would have least impact on employment generation or training of man power.

The policy of establishing a task force may have slight impact on the human resources, as it may need some additional man power and expertise in the alternate fuels and related environmental impacts. Distinct coloring scheme may have slight impact on the human resources. Awareness campaigns, training programs, and demonstrations may slightly have scope to generate additional employment opportunities. Therefore, under these criteria, all the policies are rated as shown in Fig. 7.4

Administrative Capability/Feasibility

Successful implementation of policy measures requires a well-designed system supported by appropriate infrastructure and trained personnel. In assessing the administrative feasibility of any policy, it is worthwhile to understand organizational and legal implications. It should be made known whether the current organization has the capacity and system to implement the policy. If not, would it be practical to set up a new organization or is it only necessary to improve the capacity of the current personnel through training and formal education? Do the existing

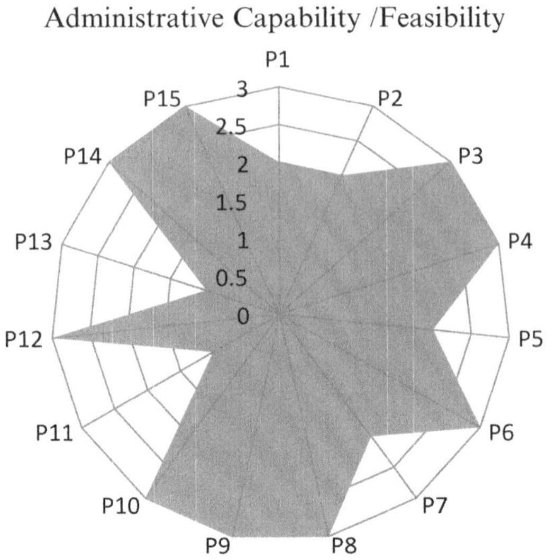

Fig. 7.5 Performance of policy measures on the criteria of administrative capability

legislations, programs, bills, and laws promote or hinder implementation of a newly crafted policy?

Policy measures like the formulation of implementing agency would call for much administrative capability. Any such drive needs extensive administrative efforts in creating and earmarking human and financial resources. Establishing partnership among various key stakeholders and providing required infrastructure essentially need additional efforts on administrative grounds. Establishment of Indraprastha Gal Limited (IGL) for the implementation of CNG for public transport in Delhi is such an example where substantial administrative efforts went in at a level of head of the state.

Incentive-based policies such as subsidies and other incentives would essentially need administrative backup. To bring CNG commercial vehicles under permit and tariff jurisdiction of the city/state Government, it was essential to amend the Motor Vehicles Act in India. This kind of administrative bottlenecks make these policies difficult ones to implement. Taxing vehicles would essentially face a similar limitation, if attempted. Providing single window/priority checks for AFVs should be relatively less cumbersome. Policies such as distinct color scheme, awareness rising, demonstrations, and trainings should be relatively easy to implement as far as administrative feasibility is concerned. However, policies such as time-bound waivers on import duty and regulation to bring in change of fuel usage essentially need substantial efforts. Figure 7.5 presents the rating of various policies on the basis of administrative capability to bring in these policies.

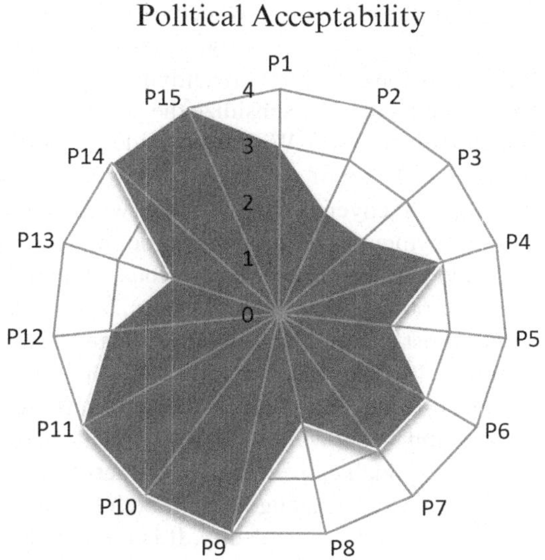

Fig. 7.6 Performance of policy measures on the criteria of political acceptability

Political Acceptability

Though transport department is a state department, it has links with environmental issues, with further links at national and international levels. Hence, these policies, though local in nature, have been greatly influenced by political agendas of the country and the leaders. In cities such as Delhi and Mumbai, urban transport and the related environmental issues have become big issues in the election campaign, and leaders have started counting on these policies to improve their chances of winning elections. Thus, it is essential for any policy to be successful; it should be politically acceptable. It may be difficult to get political support for most of the GHG reduction policies and measures because the policy makers are more likely to have their priority on local pollution control rather than giving attention to environmental issues of global concern. Hence, certain policies, which may not fit into the agenda of the set of policy makers, may be difficult to be implemented even if they have potential in controlling certain transport- and environmental-related problems. Hence, political acceptability is one of the important criteria to evaluate alternative policy measures (Fig. 7.6).

Equity Considerations

The impact of policies and measures on income distribution is equally important and therefore needs careful evaluation. These alternative fuels target mostly the public transport. This essentially reaches the lower strata of people who rely on public transport for their daily commuting. Subsidy on buses will control the possible hike in fares. This directly reaches the urban poor as a benefit.

Auto-rickshaws (three-wheelers) cater a major share of travel needs, especially as feeders to the bus and train networks. Auto-drivers are typically urban poor. Policies to convert their conventional fueled vehicles to clean fuels would hit them badly. But a policy to subsidize the additional expenditure on the conversion would add to their benefits. Without subsidy the additional expenditure can be retrieved within 30,000 km of service. With say 50 % subsidy on the additional expenditure, it could be recovered in 15,000 km, which on an average is covered in a year.

Employment generation due to policies such as formulation of an implementing body and partnership between various stakeholders would meet the urban poor to an extent of 60–75 % as the remaining 25 % would be executives.

Fuel cost difference between petrol and CNG is almost 100 % with petrol costing about 31.49 rupees a liter and CNG 16.83 per kg of gas. Hence, a regulation to convert the conventional vehicles to cleaner fuels would give the urban poor good returns, and also it would add to the state's income due to increased share of CNG in the market. Such conversions would result in substantial improvements in ambient air quality which is a huge benefit both for the state and urban poor who suffer from severe air pollution in the city. It is possible to control particulates CO and SO_x to an extent of 30–40 %.

Table 7.9 summarizes rating of various policies under different criteria and calculation of total weighted average scores and percentage average scores.

Any policy will have its effects on various domains. Hence, it is possible that one particular policy can influence/remove more than one barrier (barrier to the adoption of AFVs). Therefore, the potential of policies in removing barrier needs to be assessed based on not only certain criteria but also their potential to remove more than one barrier. Table 7.6 presents the set of barriers that each policy measure (PM) can affect. The scores derived based on normalization are applied to the percentage total weighted average score to determine the final ratings of these PAMs in controlling barriers. Table 7.10 presents the ranking of PAMs against a set of criteria alone and also the final ranking with due consideration to their multiple barrier removal capabilities.

Partnership among major stakeholders (P2) and awareness campaigns for the drivers (P15) showed highest potential in removing barriers for the adoption of cleaner alternatives and ranked first in the final ranking. Financial incentives (P4), task force to carry out checks (P7), heavy fines on defaulters (P8), demonstration of AFVs and their advantages (P11), formulation of an implementing agency to provide the necessary infrastructure (P1), and distinct coloring scheme (P9) showed better potential and are ranked among top six. Ranking of various policy measures based on only a set of criteria and the final rankings considering the potential of PMs in handling multiple barriers is presented in Fig. 7.7.

7.2.3 Measures to Promote Alternative Transport Options

According to the set of criteria and the respective analysis and assessment, distinct coloring scheme for AFVs (P9) stood first followed by awareness campaigns for the

Table 7.9 Criteria-policy matrix

		Policy alternatives							
Criteria	Score	P1	P2	P3	P4	P5	P6	P7	P8
Administrative cost	0.1	1.0	2.0	3.0	3.0	2.0	3.0	3.0	2.0
Financial burden	0.2	3.0	3.0	1.0	2.0	4.0	4.0	4.0	4.0
Human resource benefits	0.1	3.0	4.0	1.0	3.0	1.0	1.0	2.0	1.0
Administrative capability	0.3	2.0	2.0	3.0	3.0	2.0	3.0	2.0	3.0
Political acceptability	0.3	3.0	2.0	2.0	3.0	2.0	3.0	3.0	2.0
Total weighted average score	1	***2.5***	***2.4***	***2.1***	***2.8***	***2.3***	***3***	***2.8***	***2.6***
Percentage total weighted average score		**62.50**	**60**	**52.50**	**70.00**	**57.50**	**75.00**	**70.00**	**65.00**

		Policy alternatives						
Criteria	Score	P9	P10	P11	P12	P13	P14	P15
Administrative cost	0.1	4.0	2.0	2.0	2.0	2.0	3.0	3.0
Financial burden	0.2	4.0	3.0	3.0	2.0	3.0	3.0	3.0
Human resource benefits	0.1	2.0	2.0	2.5	1.0	2.0	1.5	2.0
Administrative capability	0.3	3.0	3.0	1.0	3.0	1.0	3.0	3.0
Political acceptability	0.3	4.0	4.0	4.0	3.0	2.0	4.0	4.0
Total weighted average score	1	***3.5***	***3.1***	***2.55***	***2.5***	***1.9***	***3.15***	***3.2***
Percentage total weighted average score		**87.50**	**77.50**	**63.75**	**62.50**	**47.50**	**78.75**	**80.00**

drivers (P15), training programs for the workers (P14), awareness campaigns (P10), single window/priority checkpoints (P6), financial incentives (P4), and task force to carry out check (P7) in the top six ranks. Policy measures like partnership among various stakeholders to provide infrastructure (P2), heavy fines on defaulters (P8), demonstration of AFVs (P11), and formulation of an implementing agency (P1) fall short compared to the other PMs and occupy lower ranks.

Analysis that considered the potential of PMs in addressing more than one barrier changed the ranking of the policy measures. Partnership among major stakeholders (P2) and awareness campaigns for the drivers (P15) showed highest potential in removing barriers for the adoption of CEETs and ranked first in the final ranking. Financial incentives (P4), task force to carry out checks (P7), heavy fines on defaulters (P8), demonstration of AFVs and their advantages (P11), formulation of an implementing agency to provide the necessary infrastructure (P1), and distinct coloring scheme (P9) showed better potential and ranking among the top six.

Table 7.10 Weighted scores and final rating for policy measures

Policy	Total weighted average score	Percentage total weighted average score[a]	Normalized weights of policies[b]	Final rating[c]	Percentage final scores
Formulation of an implementing agency to provide the necessary infrastructure (**P1**)	2.5	62.5 (***IX***)	0.75 (3)	1.87 (***V***)	46.87
Partnership between the government, public sector undertakings, and private actors in providing the infrastructure (**P2**)	2.4	60 (***X***)	1 (4)	2.4 (***I***)	60
Subsidies on the additional expenditure (**P3**)	2.1	52.5 (***XII***)	0.5 (2)	1.05 (***XI***)	26.25
Financial incentives like free or priority parking, separate lanes for alternative fuel vehicles, and free inspection and maintenance (**P4**)	2.8	70 (***VI***)	0.75 (3)	2.1 (***II***)	52.5
Tax on polluting vehicles and earmarking such revenue to provide subsidies for AFVs (**P5**)	2.3	57.5 (***XI***)	0.75 (3)	1.72 (***VII***)	43.12
Single window/priority checkpoints (P6)	3	75 (***V***)	0.5 (2)	1.5 (***IX***)	37.5
Task force to carry out checks (**P7**)	2.8	70 (***VI***)	0.75 (3)	2.1 (***II***)	52.5
Heavier fines on defaulters (**P8**)	2.6	65 (***VII***)	0.75 (3)	1.95 (***III***)	48.75
Coloring scheme for easy tracking of vehicles (**P9**)	3.5	87.5 (***I***)	0.5 (2)	1.75 (***VI***)	43.75
Awareness campaigns (**P10**)	3.1	77.5 (***IV***)	0.5 (2)	1.55 (***VIII***)	38.75
Demonstration of AFVs and their advantages (**P11**)	2.55	63.75 (***VIII***)	0.75 (3)	1.91 (***IV***)	47.81
Time-bound waivers on import duty (**P12**)	2.5	62.5 (***IX***)	0.25 (1)	0.62 (***XIII***)	15.62
Regulation (**P13**)	1.9	47.5 (***XIII***)	0.75 (3)	1.42 (***X***)	35.62
Training programs for the workers (**P14**)	3.15	78.75 (***III***)	0.25 (1)	0.78 (***XII***)	19.68
Awareness campaigns for the drivers (**P15**)	3.2	80 (***II***)	0.75 (3)	2.4 (***I***)	60

[a]Figure in parenthesis indicates the rank of policies according to the set criteria
[b]Figure in parenthesis indicates number of barriers removed/reduced by that particular policy
[c]Figure in parenthesis indicates final ranking

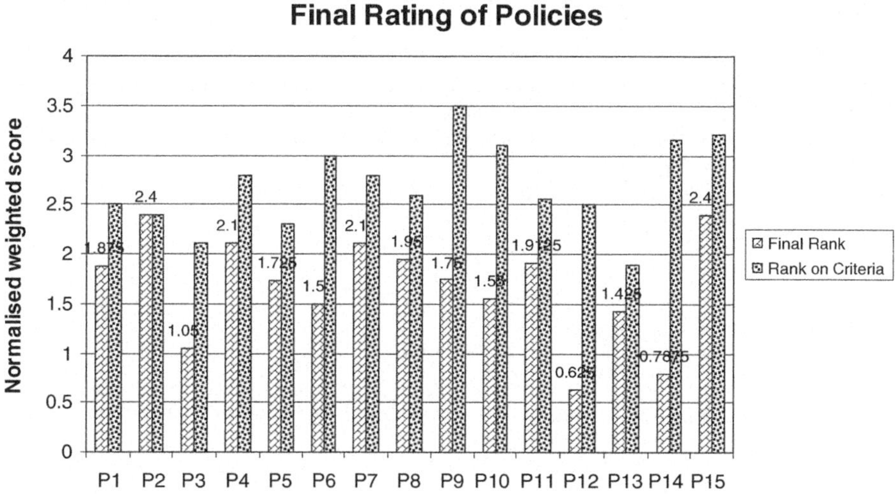

Fig. 7.7 Ranking of various policies on a set of criteria and their potential to remove number of barriers

It is interesting to see that the policy to have partnership among stakeholders to provide better infrastructure could get only tenth rank based on a set of criteria, but with its potential to handle many barriers, it could top the rankings. The top-ranked PM based on a set of criteria, color coding (P9) slipped to sixth with its limited potential to handle more than one barrier. Policy to impose heavy fines on defaulters also has its rank shifted based on the criteria. However, financial incentives (P4), task force to carry out checks (P7), color coding (P9), and awareness campaigns for the drivers (P15) were ranked high under both the criteria that testify to their potential in removing barriers. Hence the following PMs can be considered, based on both patterns of ranking, as potential PMs for the removal of barriers to cleaner alternatives (7 out of 15 PMs):

- Partnership between the Government, public sector undertakings, and private actors in proving better infrastructure (P2)
- Financial incentives like free or priority parking, separate lanes for alternative fuel vehicles, and free inspection and maintenance (P4)
- Task force to carry our checks (P7)
- Heavy fines on defaulters (P8)
- Distinct color coding for AVFs (P9)
- Demonstration of AFVs and their advantages (P11)
- Awareness campaigns for drivers (P15)

Choosing one or a combination of policy measures helps in reducing certain barriers. However, the above set of PMs would be able to control all seven pre-identified barriers to the adoption of cleaner alternatives in the transportation systems of metropolitan cities such as Delhi and Mumbai.

Reference

Yedla S (2007) Non-motorized modes of transportation for sustainable mobility: strategies for its adaptation in Mumbai, India. In: Air pollution control in the transportation sector: third phase research report of Urban Environmental Management Project. Institute for Global Environmental Strategies, Japan, pp 21–82

Chapter 8
Economic and Regulatory Measures for Environmentally Sustainable Urban Transportation

Transport sector caters for all sections of the society where the needs and expectations are widely different. The service provision in such cases is as difficult as regulating it. While the state is responsible for the provision of "transport" service, it also plays a role of regulator to control the patterns of "personalized" modes of transport. Such duel role is hard to play for the city administration in concordance with the provincial and federal governments, particularly when the degree of civic participation is marginal and the limited "space" is hardly fought for between contrasting "survival instincts." Such a situation needs not only a strong regulatory regime but also imposing economic and market mechanisms in place. Fittingly, urban transportation in India is subjected to a large set of regulatory and economic measures to provide the service with tabs on its use and strive toward sustainability. This chapter presents such regulatory and market regimes toward NMT as well as cleaner modes of transport. It presents a framework to integrate the environmental concerns into the future urban transportation planning.

8.1 Policy Measures (PMs) for the Promotion of NMT

8.1.1 Institutionalization Policies

The full potential of NMT is best achieved through the development of comprehensive strategies and policies that involve all relevant levels of government with participation from all stakeholders in NMT (Guitink 1996). As discussed in Chap. 7, *policies to integrate NMT with the public transport system planning at city/municipality level* are an important move toward implementing NMT in metropolitan cities.

The Netherlands has demonstrated the integration of NMT into mainstream transportation system. At the regional and local planning levels, it was recognized

S. Yedla, *Urban Transportation and the Environment*,
DOI 10.1007/978-81-322-2313-9_8

that adequate NMT provisions improve accessibility to centers of economic and social activity and can widen the range of personal mobility options. In the Netherlands proposals for NMT are being formulated as important components of comprehensive traffic and transport plans alongside car restraint measures and promotion of public transport. Integrating facilities for bicycles and pedestrians with mass transit systems can enhance public transport ridership by enlarging catchment areas. It would also result in fewer public transport stops, reduce vehicle operating costs and energy use, and reduce door-to-door travel times. In the privatization process of the Dutch railways, bicycle parking provisions are considered to be part of the infrastructure, just like tracks and platforms, which will remain under the financial responsibility of the central government. Such measures of institutionalization would enhance the penetration of NMT in large and upcoming metropolitan cities in India.

Integration of NMT with the other travel modes at local level provides much needed link between different travel modes (Guitink et al. 1994). An example from Colombia illustrates such integration and the resulting benefits. Productos Ramos S. A., an industrial bakery, was faced with an aging fleet of delivery trucks. Not only did replacement mean a large cash outlay, but the 135 trucks often operated with partial loads and had trouble parking near their destination. An NMT solution seemed impossible due to the large distances between the manufacturing plant and the retail outlets. The bakery introduced a decentralized computerized distribution system—a fleet of 50 trucks hauled products to satellite warehouses, where a fleet of 904 cargo tricycles picked up the goods for final delivery. As a result of this, *intermodal integration*, distribution costs dropped from 27 % to 8 % of total costs and employment increased substantially (Heierli 1993; Lowe 1989; Guitink et al. 1994). This kind of inter-modal integration can be applied both to passenger and freight transport in developing cities of Asia.

Efforts at national level are also equally important in achieving better adaptation of NMT. *Formulation of National Strategy for NMT as a facilitating framework for local plans* and *incorporation of NMT in national transportation policy* are identified as among the important policy measures to implement NMT in metropolitan cities of India. However, it is observed that national transportation plan has already provided a clause for NMT, but not to a required degree (MMRDA 2011).

In the Netherlands, at the national level, the bicycle was returned to the mainstream of government transport policy in the debate on environment and mobility during the 1980s. As part of an integrated traffic and transport approach, a Master Bicycle Plan has been developed, based on a policy statement that was approved by the Parliament. The main objectives of the Plan are to get more people onto bicycles and to attract more people to public transport by improving the public transport-bicycle chain and by facilitating cycling to and from railway stations. Such efforts at national level would provide basis for local action of integration.

Though the survey presented in Chap. 7 reveals that there is no need to make *policy to enhance the city/transport planner's capacity to integrate NMT into mainstream transportation system*, it is observed that *incorporation of standards for the bicyclists and pedestrians provisions in new road infrastructure design* would have a positive impact on the implementation of NMT in Indian cities.

8.1.2 Infrastructure Policies

Lack of infrastructure was found to be one of the major barriers for the implementation of NMT. As a remedial measure, policies *providing separate bicycle lanes on arterial roads* (fully integrated with motor vehicles/partially segregation/full segregation), *constructing wider pedestrian ways with cycling facility*, and *providing subways and overhead passages* at major intersections help to enhance the usage of NMT in metropolitan cities. Policies to *provide bicycle parking places* at all public transport stations and public places would help enhance the usage of NMT. As NMT is less technology intensive, p*olicies to support R&D for quality bicycles* are not considered important.

However, for successful implementation, this infrastructure planning needs to be well integrated with the mainstream transportation planning (GTZ 2005; Replogle 1991; Pendakur 2000). Otherwise, it could lead to disastrous results. For instance, China has a history of nearly a hundred years in bicycle use. In the late 1970s, bicycle policies focused on financial incentives to both the purchase and use of bicycles. As a result, bicycle ownership more than quadrupled in the 1980s. The policy was successful in many ways, but the lack of an integrated system approach caused problems for other transport modes: Public transport was confronted with a dramatic loss in ridership and the massive flow of bicycles caused congestion, impeded public transport, and accounted for a high share of traffic accidents. Now, the city managers focus on severe bicycle restrictions and even outright bans. The creation of adequate segregated facilities for bicycles and public transport modes, as well as their integration, could have alleviated many of the current problems (Guitink 1996).

An explicit NMT strategy is thus an important instrument for developing the potential of non-motorized-based mobility. An NMT strategy must include long-term objectives, steps to be taken to reach the objectives, and guidelines.

In a study conducted by Rastogi (2002), it was found that people in Mumbai preferred to have "physically separated lane" in order to shift to NMT. Their priority followed the order of physically separated lanes, Marked Lane (painted), Bikeways and priority crossing at intersections. About 6–11 % respondents showed their willing to shift to NMT under "modified facility scenario."

8.1.3 Financial Policies

Urban poor is predominant in Indian metropolitan cities with about 40 % populations living in slums (Census India 2001). Metropolitan cities enjoy higher per capita income compared to that of national as well as state averages (CMIE 2001). Thus, the affordability of metropolitan citizens is higher compared to the rest of the country. This concurs with the observation that low affordability is not a barrier for NMT in these cities. Moreover, higher affordability (affordability for motor ride) is the reason for the poor adaptation of NMT (bicycle in particular).

Financial incentives provide penetration of NMT into industrial workers. Policies to *support micro-credit system* (for urban poor) similar to the one given for villagers may not have much impact in promoting NMT in Indian cities. However, *policies to provide soft loans* and *relaxing excise duty on importing bicycles/bicycle parts* can impact the NMT implementation positively. In similar line of argument, policies to provide incentives for the bicycle manufacturing industry may not have substantial contributions in promoting NMT which could be due to the fact that the markets created take care of the supply needs.

Ownership of bicycle has strong correlation with income (Pandakur 2000; Rastogi 2002; World Bank 1995). Data from Malaysia and India suggest that bicycle ownership becomes substantial—40 % or more of households owning a bicycle—when household income levels reach about ten times the cost of a bicycle (Barwell and Legget 1986). Evidences from Nigeria show that households owning a bicycle had incomes 19 times its cost (Adebisi 1985). The distinction between the purchase of new and second-hand bicycles is an important, but often overlooked, qualification. A survey in Kenya shows that 65 % of bicycles are bought secondhand because of low incomes and lack of credit (Kaira 1985). A study of conditions in Malawi indicates that with bicycles costing about 45 % of annual household income, an average household simply cannot afford a bicycle without access to credit for which the poor are rarely eligible. However, with almost double the national per capita income, Mumbai households enjoy a different level of affordability, and thus, it could be a counteracting fact for the promotion of NMT.

Mumbai population, on an average (based on a study conducted on two important locations), spends 4–5 % of their income on transportation needs.[1] Table 8.1 presents the details of money spent on travel by an average Mumbai citizen.

8.1.4 Administrative Policies

One of the major barriers for NMT implementation apart from the lack of infrastructure is lack of regulations guiding the roles and responsibilities of NMT users and lack of legal base for NMT management. *Policies to make regulation*

[1] This excludes the capital costs in procuring vehicles.

Table 8.1 Details of trip length and cost per trip and percentage income spent on travel

	Trip length	Cost per trip	% Income spent
Total	20.065	5.23	1.96
Access leg	1.95 (9.7 %)	2.33 (44.11)	0.87
Primary leg	17.11 (85.3 %)	2.19 (41.88)	0.82
Egress leg	1.004 (5 %)	0.69 (36.49 %)	0.26

safeguarding NMT users and to *develop respect for the NMT users among the law enforcing officers (police) and the motor vehicle users* would positively influence the implementation of NMT in Indian metropolitan cities. This class of administrative policies supports the integration of NMT into the mainstream transportation system in a long run.

8.1.5 Social Policies

Lack of safety for the bicycle rider is one of the major barriers for NMT. Providing better infrastructure for NMT users addresses this barrier. However, to change the social acceptance and attitudes toward the NMT users, *policies to conduct awareness raising campaigns* are important and can influence the peoples' mind-sets to a great extant and help NMT to penetrate well into the existing transportation systems. For instance, it is easy to convince students and school-going children by engaging a popular sportsperson or an actor in the awareness raising campaign for NMT. In a study conducted in Mumbai where the willingness of respondents to shift to NMT was tested, it was found that 11.5–12.5 % of respondents are willing to shift to NMT once the environmental benefit statement is given to them. This demonstrates the impacts of awareness (environmental and other benefits) programs.

Confronted with severe congestion and the negative environmental effects of increasing car traffic, many industrialized countries (including Japan, Germany, Denmark, Canada, Netherlands, and the UK) are now giving more attention to the potential benefits and complementary role of non-motorized transport. The full benefits of integrated transport systems are best seen in the Netherlands which has actively pursued a pro-NMT strategy over the last decade (Guitink 1996).

8.2 Policy Measures to Promote Cleaner Alternatives in Transportation

There are a number of policies and measures discussed in the literature to achieve widespread adoption of cleaner alternatives. These policies and measures can be categorized as follows:

1. Market approach (also referred to as "economic instruments") taxes, subsidies, and sector reforms
2. Innovative financial mechanisms
3. Information, education, and technical assistance programs
4. Command and control measures (ranging from voluntary agreements to setting of standards)
5. Research and Development

Each of the above categories of policies is capable of addressing various problems effectively. However, it is impossible to achieve a sustainable solution by adopting them in isolation. It is widely accepted that a combination of the above measures is required to counter various barriers to cleaner alternatives. The following section presents all such policies that were tried in different cities or countries to overcome different sets of barriers to cleaner alternatives.

In the sections to follow, various policies implemented in different countries to promote cleaner technological options in urban transport system are explained in detail.

8.2.1 Market Approach

Economic instruments such as taxes, subsidies, and user fees play an important role in the transport sector. Many countries have used them in order to control the travel patterns and their associated emissions. Fuel taxation plays an important role in all the policy issues concerning the transport sector. Differentiated fuel taxes are also widely used to support or limit the use of specific fuels. However, the response to fuel tax increase is very limited in the short term although somewhat greater in the long run. In many countries, fuel taxes are already at levels that are not politically popular, and the additional increase will create problem in the future. In this section various cases of taxes, subsidies, and fees as applied in different countries are presented.

High *tax on gasoline* was used as a replacement for subsidy to increase the usage of natural gas in Argentina. Instead of offering subsidies on CNG, the Government of Argentina had adopted high tax on gasoline as an incentive mechanism to encourage fuel switching. Price difference between gasoline premium and the natural gas supplied to the industry encouraged promotion of gas stations and conversions to natural gas operated vehicles. In a similar approach to promote alternative fuels, Denmark has employed *fuel consumption tax* waivers on electric vehicles (EVs) to counter their CO_2 emissions. Fuel consumption tax is imposed on gasoline and diesel vehicles.

Taxes on emissions from large and medium sources have been used as an effective tool in Czech Republic. In Finland, carbon tax is applied to all forms of energy. The carbon tax is levied per ton of CO_2 emitted. A fuel tax was used of approximately FIM 3.00 per liter for petrol and FIM 1.50 per liter for diesel. Lower value added tax (VAT) is levied on public transport tickets (VAT applied at 6 % as against normal rate of 22 %). Taxation is designed to promote more fuel-efficient cars. *Value added tax* and acquisition tax are exempted by the Korean Government as an incentive when a CNG bus is purchased. In addition, environment improvement charges imposed on diesel vehicles are exempted for the CNG buses in Korea.

The Government of Japan provides an amount equal to 50 % of the difference in cost between electric vehicles and regular automobiles. Low-emission vehicles (CNG-fueled vehicles, hybrid engine, and methanol-fueled vehicles) attract lower automotive sales tax as well as annual automobile ownership taxes and acquisition tax. The systems that supply alternative fuels also receive favorable tax treatment.

In an attempt to use relaxation on *excise tax* to introduce cleaner fuel in transport sector, Federal Government of Canada has employed the elimination of federal excise tax at the rate of 8.5 cent per liter on the alcohol proportion of petrol-ethanol. As a result the market share of alternative fuels reached 2 % by 2000.

In a move to support the Supreme Court mandate to convert the bus fleet to CNG in Delhi, the Government of India (GoI) has exempted CNG converter kits imported into the country from customs duty. The other incentives include an exemption in the 12 % state sales tax and a 3 % low-cost loan option. An excise tax waiver for new CNG vehicles and a waiver on the customs duty of imported CNG kits from the Finance Department have resulted in a 15 % decline in the price of a converter kit. With such measures, Delhi has successfully implemented a shift of its public and quasi-public modes of transport to clearer fuels.

Denmark has used exemption of *import fee* on electric vehicles (EVs) as an effective tool to cut down CO_2 emissions from their transport sector. Electric vehicles can be imported without paying the import registration fee up to 180 % of the price of a vehicle.

In an effort to use *subsidy* as a tool to modulate transport sector and related environmental emissions, the French Government provided a subsidy of US $ 8,000 on purchase of electric vehicles and the French Electric Company provided a further subsidy of $4,000 for charging the vehicles. Japanese Government provides subsidies to local public organizations or private companies that introduce electric, natural gas, or other low-emission vehicles and systems that supply fuel for these vehicles. Direct incentives are available in Japan for hybrid passenger vehicles such as the Toyota Prius or Honda Insight. Substantially higher incentives are available for large vehicles such as buses and trucks.

In order to introduce 20,000 CNG buses and to construct 400 gas stations, the MOE and local governments in Korea propose to spend a total of USD 610 million by the year 2007. Subsidy is provided to bus companies and low interest loans offered to the gas station owners.

EVs were given *free parking* as an incentive. *Free charging* of EVs is provided at several EV-specific parking places in Denmark. US Government provides many incentives to promote electric vehicles. California State of USA has the zero emission vehicle incentive programs, which provides support toward the purchase or lease of EVs. Los Angeles Department of Water and Power provides discount for electricity used to recharge EVs during off peak time periods. Many cities in the

USA provide incentives for installation of EV charging station. Los Angeles Airport offers free parking and charging for EVs.

The US government has attempted many *innovative market tools* to control emissions from transport sector. Such major tools are discussed below:

- Tax credit for electric vehicles (EVs) and hybrid electric vehicles. Federal income tax deductions for alternative-fueled vehicles were also implemented, for example, up to US$2,000 for "clean fuel" cars, US$5,000 for light commercial vehicles, US$50,000 for heavy trucks and buses, US$100,000 for refueling stations/equipment, and up to US$4,000 for electric vehicles. A number of states also offer further vehicle tax deductions, rebates, or credits for environmentally friendly vehicles.
- A "Gas Guzzler" tax is levied at an increasing rate for cars with less fuel efficiency.
- Reform of federal tax subsidy for employer parking was introduced such that employees can choose between US$65 per month extra in taxable income or the parking space. This was designed to encourage the use of public transport and carpooling and so on for commuting.
- License plate fee reduced from $25 to $8 for alternative fuel vehicles (AFVs).
- Vehicle license tax reduced by over 95 % (worth $400–$600 the first year, less each successive year) on AFVs.
- Personal state tax credit of up to $7,500 for light-duty AFVs and up to $30,000 for heavy-duty AFVs.
- Personal state tax credit of up to $2,000 for personal AFV refueling equipment.

8.2.2 Innovative Financial Mechanisms

Transportation infrastructures such as road and metro are capital-intensive, and conventional means of finance doesn't work in countries like India. It needs augmentation by means of other innovative mechanisms.

Innovative financial mechanisms in transportation have three functions, viz., augmenting financial needs for infrastructure so that a better infrastructure is made available, controlling of automobile use, and internalizing the externalities so that the collected fund can be used to protect the damaged entity. Some are in the form of taxes and fee and others are in the form of incentives and rebates to promote environmentally friendly options.

Now there are numerous examples in the world to demonstrate that provision of better roads does not bring commensurate economic benefits and results in irreversible environmental damage. Therefore, one has to be judicious in applying these finance mechanisms.

In the Indian context, while infrastructure provision does offer great expectation, continuity of infrastructure augmentation and the service provision needs to be ensured on a long-term sustainability. Therefore, it is essential to employ innovative financial mechanisms as emphasized by the Government of India in various policy documents; participation of private sector is the key in meeting the huge investment requirements in this sector.

Financial policies and regulatory policies complement each other and should be used as a package. While regulatory policies safeguard the public welfare, conserve, and protect resources, fiscal policies generate the necessary revenues to cover the investment requirements. Fiscal policies can also capture the negative externalities and also signal the users of the damages.

Government of India has been promoting innovative financial mechanisms in the form of PPP and also working toward enhancement of capacity of city administrators in making the necessary regulatory amendments to foster such partnerships in Indian cities.

Following are the list of different types of fiscal policies applied at different levels and in different countries (Gordon 2005):

- Fuel taxes and fee
  - Gasoline and diesel taxes—Poland
  - Carbon tax—Sweden
  - Environmentally base fuel fees
- Vehicle taxes
  - Annual vehicle attribute-based taxes and fees—EU Japan and Singapore
  - Tax/fee reductions or exemptions for new clean and efficient cars—Japan, Denmark, and Germany
  - Annual, externality charges for carbon dioxide and smog—the UK and Denmark
- New vehicle rebate program
  - Rebates—Japan and the USA
  - Fees—US Gas Guzzler tax
  - Freebees—Austria
- Road fee
  - Road pricing—the USA, India, and Japan
  - Congestion pricing—London
  - Full externality-based road pricing—Singapore
- Parking pricing
  - Parking fee—California, Japan, and EU
  - In lieu fees for parking—South Africa, Iceland, Canada, Germany, the USA, and the UK
  - Replacing parking supply with programs that reduce parking demand—the USA

- Vehicle insurance fees
  - Fines for lack of mandatory insurance—the UK and the USA
  - Insurance-specific automobile taxes—France
  - Pay-as-you-drive and pay-at-the-pump insurance—the USA and the UK
- Fleet vehicle incentives
  - Cost-effective clean, fuel-efficient public, government vehicle fleets—Canada
  - Incentives to promote clean company cars—the UK
  - Incentives for environmentally based vehicle leases

It is important to choose an appropriate fiscal policy subjected to the prevailing conditions. Fiscal policies such as taxes and fees can be regressive and can impact the low-income populations more than the others.

Following are a few examples of best practices where different fiscal policies are applied mostly for the regulated use and to internalize the damage costs.

Gasoline and Diesel Taxes (Poland): The fuel tax imposed by different countries varies widely. Indexing gas taxes to the wealth of a country helps explain why car ownership and use is lower in certain countries. Eastern Europe with its high gasoline tax indexed to GDP has fewer cars compared to the USA whose indexed value is almost insignificant. Some countries may have relatively less fuel tax, but it could be very effective as it could give high index value over their GDP.

Regulatory policies have to be in place to avoid the rebound effect of the fuel tax. Vehicles running on cheaper fuels usually have a longer mileage and the usage. Therefore, the regulatory policy should be made so that the car manufacturers would control the mileage for those "dirty vehicles."

Carbon Taxes (Sweden): Carbon tax is a level II fuel tax. A tax is imposed on the carbon dioxide emitted by the vehicle. This tax affects all parts of the economy, and it can be coupled with tax reductions on labor or income. Carbon taxes are known for their permanent incentive to emissions reduction compared to the fuel taxes.

In a measure to complement the existing energy tax system, a carbon tax was introduced in Sweden in 1991. However, it has come with too many exceptions where the industrial sectors pay no energy tax and only 50 % of the general carbon tax, while electricity production sector pays neither energy nor carbon taxes.

By 1997, the carbon tax was at 0.365 SEK/kg CO_2 (approximately US$150 per ton carbon). As this rate is so low, it failed to affect consumers' demand behavior for gasoline or diesel fuels.

If the low-emission diesel fuel and high-emission diesel fuel are both taxed at the same level despite causing different levels of environmental damage, the tax would fail to bring a change. Therefore, Sweden had adopted a differential tax based on the sulfur content and that had resulted in rapid switch to ultralow sulfur diesel fuel. Sweden's carbon tax has been successful in reducing emissions of carbon dioxide.

Congestion Pricing (London): London-wide congestion pricing was discussed as part of a wider strategy, including public transport improvements, parking/loading enforcement, and traffic management. In 2003, a congestion charging zone was established in central London with a boundary of the zone formed by the Inner Ring Road. Ride on the ring road is not charged. Vehicles are given a registration number, and a flat daily charge of £5 applies between 7 a.m. and 6:30 p.m. It can be paid in the form of weekly, monthly, and yearly passes without a volume discount. Automatic license plate numbers are used to monitor them by means of fixed cameras. If payment is made after 10 p.m., the charge would be £10. If no record of payment is made by midnight, a £80 penalty is charged to the registered vehicle owner. Persistent evaders are clamped or towed. Taxis, licensed minicabs, emergency service vehicles, blue/orange badge holders, and alternative energy vehicles are exempted.

Within 6 months, the beneficial effects on traffic conditions in central London were visible with the traffic delays inside the zone having been reduced by 30 % and travel times 14 %. While some use public transport, about one-quarter now drive around the perimeter of the zone, avoiding the charge. Others have switched to other modes, such as taxis. As a result, the public transport has increased ridership, and about 60,000 fewer car trips were coming into the congestion pricing zone. London congestion pricing has been successful in decongesting and also could accrue net annual revenue in the range of £80–100 million a year.

Various *financial incentives* are provided by the Federal Government of *Canada* to encourage the development of the natural gas vehicle market. For example, $500 is provided for vehicle conversion, $1,000 toward the purchase of a new natural gas vehicle, and $50,000 for each new vehicle refueling station. This along with relaxed excise tax on cleaner fuel has resulted in increase in their market share.

Cheaper loans on alternative fuel vehicles are given as an effective instrument to promote cleaner fuels in various countries such as Thailand, Bangladesh, and India.

In India, CNG is available at a price in gasoline—equivalent terms of around Rs. 8.1–9.2/l, which compares favorably with gasoline prices of Rs. 22–26/l in 2001. It is estimated that the costs of converting a vehicle to CNG in India can be recovered after 30,000 km.

Improper fiscal policies would lead to huge financial loss to the state. In Argentina, CNG was introduced and the government failed to create a fiscal mechanism by keeping the difference between diesel price and CNG price very low. Hence, there was little conversion from diesel to CNG because the price differential between diesel and CNG was not sufficient to recover the incremental cost of NGVs within a reasonable period. As a result, there are no CNG buses in regular operation today, and in fact diesel is actively competing with CNG to wrest the taxi market from CNG.

8.2.3 *Information, Education, and Technical Assistance Programs*

Indian Government has undertaken many demonstration projects to promote the use of alternative fuels such as propane and electricity. There are also demonstration programs in place for converting three-wheelers to electrically powered buses and vehicles that operate at the Taj Mahal by using electric batteries. Although the Indian government has cleared the use of LPG, laws and regulations governing its use in vehicles have not been put in place. There are concerns over the safety of LPG conversions because of lack of overseeing of the installation of kits. An Indian company has designed an LPG conversion kit for scooters.

8.2.4 Command and Control Measures (Ranging from Voluntary Agreements to Setting Up of Standards)

Argentina has adapted a command and control approach by developing standards for a successful national private industry of compressors and dispensers, cylinders, and conversion kits to promote natural gas usage in transport sector. Certification policies were implemented by the National Gas Regulatory Agency and carried out by internationally well-known Certification Agencies.

Mandatory *vehicle inspections* with minimum emissions requirements and zero lead standards were implemented in Finland. Investment in electric railways is done through the Finland Government. Privatization of railways is adopted to improve efficiency and development of high-speed passenger train connections to make public transport more attractive.

French government regulations of urban transport include requirements that some public entities managing fleets of more than 20 vehicles must purchase "clean" vehicles representing 20 % of the total vehicle purchase. The law on Air and Rational Use Energy provides funds for costs related to choosing, buying, and using alternative fuel vehicles such as reimbursement of 50–70 % of the cost of fleet orientation and diagnostic studies that assist in the choice of vehicle and fuel systems; funding for the acquisition of alternative fuel vehicles for use in demonstration programs; incentives that narrow the difference in prices between alternative and conventional fuel vehicles; and provisions for reimbursement of taxes on alternative fuels under some circumstances and other tax credits.

In order to reduce pollution in the City of Milan, the Italian government only permits electric vehicles, bicycles, and motorcycles fitted with catalytic converters in the streets between 8:00 a.m. and 8:00 p.m. The cost of LPG is around 50 % lower than the cost of gasoline.

The popular "*Delhi Decision*" has set the command and control approach to environmental management in a different league. As a landmark decision, the Supreme Court of India ruling went into effect in New Delhi on March 31, 2001, that mandated the conversion of the entire bus fleet to compressed natural gas (CNG). In addition, the honorable court has ordered the replacement of taxis and auto rickshaws with engines running on clean fuels. Recently, there was a Supreme Court order to the effect that government bodies in Delhi should convert their petrol-driven cars to CNG cars. Though it has created considerable friction in the system in the short run, it has resulted in rich dividends in terms of improved air quality in the long run.

8.3 Review of Policy Initiative and Regulatory Measures in Indian Transport Sector

8.3.1 Emission Standards and Regulatory Measures

Standards for controlling emission levels of new vehicles were incorporated into the Motor Vehicles Act (1989) as late as 1991 to enforce some quality control on the automobile industry. The mass emission standards refer to grams of pollutant emitted per kilometer traveled by the vehicles during mass emission tests conducted under specified driving conditions, as notified by the Ministry of Surface Transport and enforced by State Transport Departments.

The existing Bureau of Indian Standards (BIS) for fuel started incorporating emission parameters during the same period. But in both cases, the standards were set according to the terms and conditions dictated by the industry. Both the regulatory authorities and the manufacturers pleaded incapacity to make any drastic improvements. The standards they followed were far too lax compared to that elsewhere in the world. In 1996, the government came up with mass emission standards for vehicles that are stricter than earlier. Other measures during 1994–1996 include the use of unleaded petrol and fitting of catalytic converters in the cars. Fitting catalytic converters to cars have reduced emissions of CO. Now EURO II, EURO III, and EURO IV norms have also been enforced.

Poor maintenance of vehicles adversely affects their emission efficiency. The role of maintenance in combating vehicular pollution was reflected in government policy for the first time in 1989, which made the certificate of fitness mandatory for registration of public vehicles, commercial vehicles, and personal vehicles older than 15 years. The 1990 vehicular emission rules required all motor vehicles to comply with the exhaust emission standards as laid down. Delhi State Transport Department issues Pollution Under Control (PUC) certificates to vehicles. Vehicle owners are required to check the emission level of their vehicles every 3 months and obtain a PUC certificate. Vehicles that fail to meet the standards are required to rectify the fault and obtain the certificate. The State Transport Authority fines vehicles that do not possess a PUC certificate.

This is a step toward minimizing vehicular pollution by regular checks. This system, however, has come under severe public criticism owing to the existing lacunae in the issue of certificates and the discrepancies in pollution readings from one station to another. For example, though it is necessary for all vehicles to have a valid PUC at all times, according to statistics maintained by transport department, percentage of vehicles with valid PUC was highest at 23 % in 1997 since the introduction of this legislation. There is a provision to levy a monetary fine of Rs. 1,000 on motorists who fail to abide by the law. However, enforcement of the law is poor. For instance, since the introduction of the law in 1991, vehicles fined, as a percent of those without PUC was highest at 1.07 in 1997. Statistics maintained by Automobile Association of Upper India reveals that more than 50 % of vehicles in Delhi in May 1995 failed to comply with the prescribed standards. What is more alarming is that nearly 44 % of the new vehicles checked were found to be not in compliance with the standards. This shows that PUC despite being a potentially powerful instrument in controlling pollution from vehicles has failed to make an impact on vehicular pollution. This, apart from cutting down the emissions, actually created a lot of awareness among citizens for the environmental pollution.

8.3.2 Judicial Interventions

The failure of the administration to enforce environmental regulations has led to judicial interventions. The Supreme Court has come up with several guidelines in the last few years. The Court has urged the government to accept the emissions standards EURO I, II, III, etc., for the vehicles as adopted by the European Commission. In the last few years, the Supreme Court has issued a number of directives that aim at environmental emission controls. Such directives are discussed in Chap. 2 (CPCB 1999).

In another demonstration of legal institutions taking initiatives for improvement of environment, Mumbai High Court set up a Committee on December 1999, headed by the Transport Commissioner, to examine the entire matter and to come up with future directions. Salient features of the same are discussed in Chap. 2.

8.3.3 Measures to Promote Cleaner Fuels

Two of the major initiatives toward emission control are enforcement of unleaded petrol first in mega cities and then in the entire country and low sulfur diesel. These initiatives implemented in April 1, 2000 and 1999, respectively, have come out with a visible impact on deteriorating environmental quality especially in Delhi.

Unleaded petrol: The specification of lead in Indian petrol used to be 0.56 g/l max. in 1994. Lead has been phased out from petrol during the last 6 years in India. From February 1, 2000, only unleaded petrol is produced and sold in the entire country. In the developed countries, lead phasing out was spread over a period of 10–20 years. China and many other countries in the Asia Pacific region and in Europe, South America, and Africa have not phased out lead as yet.

Reduction of sulfur content: The sulfur content in diesel has been reduced by 75 % from 1.0 % max. in 1996 to 0.25 % max. in the entire country in a period of 4 years from April 1, 1996, to January 1, 2000. The government had approved setting up of nine diesel hydro desulfurization (DHDS) plants in nine refineries for reducing the diesel sulfur content from 1.0 % max. to 0.25 % max. at a total cost of Rs. 5568.31 crores (1 crore = 10 million) in June 1997. This has enabled supply of diesel with 0.25 % max. sulfur in the entire country from January 1, 2000. Further, in the four metro towns, sulfur content in diesel has been reduced by 95–0.05 % max. The entire unleaded petrol and low sulfur program are discussed in Chap. 2

(CPCB 1999). With these initiatives, reduction in vehicular emission load and improvement in ambient air quality are effortlessly visible in Delhi.

8.4 National Urban Transportation Policy

NUTP is the first country-level policy providing guidelines for the integrated planning of transportation system in Indian cities. The objective of this policy is to ensure safe, affordable, quick, comfortable, reliable, and sustainable access for the growing number of city residents to jobs, education, recreation, and such other needs within our cities. This is sought to be achieved by:

- Incorporating urban transportation as an important parameter at the urban planning stage rather than being a consequential requirement
- Encouraging integrated land use and transport planning in all cities so that travel distances are minimized and access to livelihoods, education, and other social needs, especially for the marginal segments of the urban population, is improved
- Improving access of business to markets and the various factors of production
- Bringing about a more equitable allocation of road space with people, rather than vehicles, as its main focus
- Encouraging greater use of public transport and non-motorized modes by offering central financial assistance for this purpose
- Enabling the establishment of quality-focused multimodal public transport systems that are well integrated, providing seamless travel across modes
- Establishing effective regulatory and enforcement mechanisms that allow a level playing field for all operators of transport services and enhanced safety for the transport system users
- Establishing institutional mechanisms for enhanced coordination in the planning and management of transport systems
- Introducing Intelligent Transport Systems for traffic management
- Addressing concerns of road safety and trauma response
- Reducing pollution levels through changes in traveling practices, better enforcement, stricter norms, technological improvements, etc.
- Building capacity (institutional and man power) to plan for sustainable urban transport and establishing knowledge management system that would service the needs of all urban transport professionals, such as planners, researchers, teachers, students, etc.
- Promoting the use of cleaner technologies
- Raising finances, through innovative mechanisms that tap land as a resource, for investments in urban transport infrastructure
- Associating the private sector in activities where their strengths can be beneficially tapped
- Taking up pilot projects that demonstrate the potential of possible best practices in sustainable urban transport

Ministry of Urban Development is largely responsible for the provision of clean and safe transport to urban areas, and the efforts in these lines are in the form of improving public transport and promoting non-motorized modes (NMT) of transport. The National Urban Transport Policy launched by the ministry provides the basic guideline for the development of infrastructure and also for service provision. No specific policy is being made for urban poor as they are the captive users of public transport and NMTs.

8.5 Internalizing Environmental Concerns in to the Long-Term Urban Transportation Planning

Different alternative options show different potential to save energy and to control local and global emissions. They also show different potential on economic performances. Long-term transportation planning to augment the transportation fleet, in order to meet the travel needs of a city, needs to consider these alternative options and find an optimal mix of vehicles. Such an attempt to find the optimal mix of vehicle would also be influenced by the set of constraints toward emission control. Though there exist a number of alternative options in discrete attempts, a least cost scenario needs to be developed to achieve the control in pollution over a period of time and at a least cost. For this the above list of alternative options needs to be optimized to achieve a globally optimal solution. This section presents dynamic optimization model-based methodology to find an optimal mix of vehicular fleet with multiple objectives of meeting the travel needs and emission cuts. Various scenarios of emission control are developed with varying emission cuts to identify suitable vehicular mix.

8.5.1 Least Cost Approach to Long-Term Urban Transportation Planning: Optimization Model

8.5.1.1 Methodology

The total cost includes capital cost and operational and maintenance cost of the vehicles that should be added during the planning horizon and the operational and maintenance cost of the existing vehicles for the passenger transportation. All costs are expressed as a total net present value to the base year (Yedla et al. 2005).

- **Variable**

X_{idv} Number of vehicle, mode *i* device *d* to be commissioned in year *v*

V_{idvt} Km traveled by vehicle mode *i*, device *d*, vintage *v*, year *t*

- **Parameters**

C_{idv}	Discounted capital cost of a vehicle mode *i*, device *d*, vintage *v*
S_{idv}	Discounted salvage value of a vehicle mode *i*, device *d*, vintage *v*
O_{idvt}	Operating cost of vehicle mode *i*, device *d*, vintage *v*, year *t*,
OC_{idt}	Occupancy rate of a vehicle, mode *i*, device *d, year t*
L_{idv}	Life period of a vehicle, mode *i* device *d,* vintage *v*
LF_{idt}	Load factor of a vehicle, mode *i*, device *d*, in year *t*
FA_{jt}	Fuel availability of fuel type *j* in year *t*
SFC_{idvt}	Specific fuel consumption of a vehicle, mode i and device d, vintage *v*, year *t*
TD_t	Total travel demand in p-km in year *t*
$\mathrm{TD}_{it}^{\max}$	Maximum level of total travel services by transport mode *i* in year *t*
$\mathrm{TD}_{it}^{\min}$	Minimum levels of total travel services by transport mode *i* in year *t*
$V_{idt}^{\max}$	Maximum km traveled by a vehicle, mode *i*, device *d*, year *t*
$X_{idt}^{\max}$	Maximum number of candidate vehicle, mode *i*, device *d*, year *t*
EF_{idvt}	Emission factor of vehicle, mode *i* device *d*, vintage *v*, year *t*
$E_t^{\max}$	Target level of CO_2 emission in year *t* from all vehicles
$E^{\max}$	Target level of overall CO_2 emission during the planning horizon
$\mathrm{GWP}_t^{\max}$	Target level of global warming potential in year *t*
$\mathrm{GWP}^{\max}$	Target level of global warming potential during the planning horizon

Objective Function

To minimize total costs (capital, operational, and maintenance cost) of new vehicles and operating and maintenance costs of existing as well as new vehicles.

$$\sum_{i=1}^{I}\sum_{d=1}^{D}\sum_{v=1}^{V} X_{idv}(C_{idv} - S_{idv}) + \sum_{i=1}^{I}\sum_{d=1}^{D}\sum_{v=-v}^{t}\sum_{t=1}^{T} V_{idvt}*O_{idvt}$$

8.5.1.2 Travel Demand Constraints

Passenger Travel Demand

1. Transport services supply in *p*-km in year *t* must be greater than the forecasted demand in the year *t*. *p*-km is calculated by multiplying the *v*-km by occupancy rate:

$$\left(\sum_{i=1}^{I}\sum_{d=1}^{D}\sum_{v=-v}^{t} V_{idvt}\right)*\mathrm{OC}_{idt} \geq \mathrm{TD_t} \quad \forall_t$$

2. Total travel services in *p*-km by each mode (bus, car, 2-wheeler, and 3-wheeler) of transport services in year *t* must be greater than that of the minimum level in

year t. When the share of HOV increased, the minimum demand for the low-occupancy vehicles (car, 3-wheeler and 2-wheeler) will be reduced from the forecasted demand but the minimum demand for the HOV will not be reduced:

$$\left(\sum_{d=1}^{D}\sum_{v=-v}^{t} V_{idvt}\right) * \mathrm{OC}_{idt} \geq \mathrm{TD}_{it}^{\min} \quad \forall_{i,t}$$

Freight Travel Demand

3. Transport services supply in ton-km in year t must be greater than the forecasted demand in the year t. *ton-km* is calculated by multiplying the v-km by load factor (only for goods vehicles):

$$\left(\sum_{i=1}^{I}\sum_{d=1}^{D}\sum_{v=-v}^{t} V_{idvt}\right) * \mathrm{LF}_{idt} \geq \mathrm{FD}_t \quad \forall_t$$

4. Total travel services in ton-km by each mode of transport services in year t must be greater than that of the minimum level in year t:

$$\left(\sum_{d=1}^{D}\sum_{v=-v}^{t} V_{idvt}\right) * \mathrm{LF}_{idt} \geq \mathrm{FD}_{it}^{\min} \quad \forall_{i,t}$$

Availability

Here, the number of vehicles is related with vehicle utilization. Total km traveled by each type of vehicle cannot exceed its availability, which depends on the vehicle utilization rate. The maximum vehicle-km that can be traveled by a vehicle could be the average vehicle utilization rate:

$$\left(\sum_{v=-v}^{t} V_{idvt}\right) \leq \left(\sum_{v=-v}^{t} X_{idv} - \sum_{t=1}^{T} \mathrm{RV}_{idt}\right) * V_{idt}^{\max}$$

Total retired number of vehicles can be represented by the following equation:

$$\sum_{t=1}^{T} \mathrm{RV}_{idt} = \sum_{v=-v}^{t-L_{idv}} X_{idv}$$

Then the availability constraint can be written as

$$\left(\sum_{v=-v}^{t} V_{idvt}\right) \leq \left(\sum_{v=-v}^{t} X_{idv} - \sum_{v=-v}^{t-L_{idv}} X_{idv}\right) * V_{idt}^{\max} \quad \forall_{i,d}$$

Vehicle Stock

1. Maximum vehicle stock: for the new technology, total number of new vehicles cannot exceed the total number of vehicles that can be added during the planning horizon (depends on penetration rate):

$$X_{idvt} \leq X_{idt}^{\max} \quad \forall_{i,d,t} \text{ (For some selected new technology)}$$

Resource Constraints

For some selected fuel, the resource constraint can be applied. A total travel service in v-km depends on the fuel availability for some selected fuels:

$$\sum_{v=-v}^{t} V_{idvt} * \mathrm{SFC}_{idvt} \leq \mathrm{FA}_{jt} \quad \forall_{j,t}$$

Emission Constraints

This constraint is used when the target is CO_2 emission reductions. Emission constraints: the CO_2 emissions are calculated by multiplying the vehicle-km by the CO_2 emission factor.

1. Annual emission constraints: total carbon dioxide emissions by all types of vehicles in a year should not exceed the target level of emission of that year:

$$\sum_{i=1}^{I}\sum_{d=1}^{D}\sum_{v=-v}^{t} V_{idvt} * \mathrm{EF}_{idt} \leq E_t^{\max} \quad \forall_t$$

2. Overall emission constraints: total carbon dioxide emissions by all types of vehicles during the planning horizon should not exceed the target level, depending on overall emission reduction:

$$\sum_{i=1}^{I}\sum_{d=1}^{D}\sum_{v=-v}^{t}\sum_{t=1}^{T} V_{idvt} * \mathrm{EF}_{idt} \leq E^{\max}$$

This constraint is used when the target is global warming potential (GWP). The most recent GWPs (assigned in IPCC 1996) for the most greenhouse gases are used for the estimation of GWP. GWPS are not provided for the criteria pollutants CO, NO_x, NMVOCs, and SO_2 because there is no agreed-upon method to estimate the contribution of gases that have only indirect effects on radiate forcing (IPCC 1996).

GWP: The GWP is estimated by multiplying the vehicle-km by the emission factor and with respective GWP assigned in 1996.

3. Annual GWP constraints: total GWP by all types of vehicles in a year should not exceed the target level of GWP of that year:

$$\sum_{i=1}^{I}\sum_{d=1}^{D}\sum_{v=-v}^{t} V_{idvt}\left\{1*(\mathrm{EF}_{idt})_{\mathrm{CO_2}} + 21*(\mathrm{EF}_{idt})_{\mathrm{methane}} + 310*(\mathrm{EF}_{idt})_{\mathrm{N_2O}}\right\}$$
$$\leq \mathrm{GWP}_t^{\max} \quad \forall_t$$

4. Overall GWP constraints: total GWP by all types of vehicles during the planning horizon should not exceed the target level, depending on overall GWP reduction:

$$\sum_{i=1}^{I}\sum_{d=1}^{D}\sum_{v=-v}^{t}\sum_{t=1}^{T} V_{idvt}\left\{1*(\mathrm{EF}_{idt})_{\mathrm{CO_2}} + 21*(\mathrm{EF}_{idt})_{\mathrm{methane}} + 310*(\mathrm{EF}_{idt})_{n_2o}\right\}$$
$$\leq \mathrm{GWP}^{\max}$$

This model can be applied to any city by selecting a number of alternative transport options. This model needs the following information:

- List of alternative options
- Energy consumption of the alternatives
- Emission factors of all pollutants for each alternative
- Cost of each alternative in the form of per unit cost of operation and maintenance
- Capital cost of alternatives
- Travel demand for the entire planning period

Scenarios based on targeted emission reductions can be developed, and accordingly the optimization model presents the following outputs:

- Mix of vehicles, old and new alternatives, over the planning period
- PKM catered by different modes over time line
- Total cost of transportation
- Total environmental emissions
- Total carbon dioxide emissions

The optimization model successfully attempts to plan transportation system for the planning period by aiming to cut certain emissions and also minimize the costs for future years. This model can help to institutionalize the integration of environmental sustainability into transportation planning.

In the present case study, freight travel demand constraint is not considered, as there is no room for the freight transport in the list of potential candidates for reducing pollution. The above analysis is aimed at determining the optimal mix of vehicle in any particular year over a time period to achieve the least cost of overall transportation with a set of constraints. Marginal abatement cost of each pollutant under different GHG mitigation scenarios is determined.

8.5.2 Case Study

The technical and economic feasibility analysis carried out for the selected alternatives as presented in Chap. 4 is used in the optimization exercise to determine the optimal mix of vehicle stock for Mumbai transport system.

Alternatives considered are as follows:

- CNG 3-wheelers in place of conventional 3-wheelers
- Battery-operated 3-wheelers in place of conventional 3-wheelers
- CNG cars in place of conventional fuel cars (this includes personalized cars and taxis)
- CNG buses in place of conventional fuel buses

Least-cost approach was adapted to determine optimal mix of vehicles for different scenarios. Based on GHG mitigation targets, the following scenarios are developed and applied in the case study.

Business-as-usual scenario: Base case without emission regulation

Mitigation scenario I (MS-I)	5 % reduction in the total CO_2 emissions over the time period under consideration
Mitigation scenario II (MS-II)	10 % reduction in the total CO_2 emissions over the time period under consideration
Mitigation scenario III (MS-III)	15 % reduction in the total CO_2 emissions over the time period under consideration
Mitigation scenario IV (MS-IV)	20 % reduction in the total CO_2 emissions over the time period under consideration
Mitigation scenario V (MS-V)	25 % reduction in the total CO_2 emissions over the time period under consideration

Optimization model was run to determine the total cost of transportation for Mumbai transport system. Emission of CO_2 constitutes a set of constraints in this optimization model. BAU scenario considers no emission constraint. Twenty years (1997–2020) was chosen as the planning horizon.

The model output is the vehicular mix (old and new alternatives) that can meet the travel demands that have been calculated for the case study. Table 8.2 present travel demand and the model output of supply which are in balance.

Total least cost for Mumbai transport under BAU scenario was found to be 4,276,596 USD. The total cost of Mumbai transportation for mitigation of 5 % CO_2 and 25 % CO_2 emission increased by 0.497 % and 1.99 %, respectively. Total cost of Mumbai transport system is presented in Table 8.3.

Vehicular augmentation for the future transport planning in Mumbai, CNG 3-wheelers, BOV 3-wheelers, CNG cars, and CNG buses is more prominent. The

Table 8.2 Travel demand and supply for different modes of transport in Mumbai

	Travel demand (million PKM)			Supply (million PKM)		
	3-W	Cars	Buses	2-W	Cars	Buses
1998	3,600	3,400	21,400	3,600	5,400	21,400
2005	6,000	8,300	43,300	6,000	8,299.9	43,400
2010	6,800	10,800	56,700	6,800	10,800	56,700
2015	9,700	14,300	67,900	9,700	14,300	68,000
2020	13,800	18,700	92,700	13,800	18,700	93,200

Table 8.3 Cost for CO_2 reduction under different scenarios for Mumbai

Emission reduction level (%)	Total cost (USD)
5	4,297,872
15	4,302,128
20	4,319,149
25	4,340,426
30	4,361,702

change in vehicular mix due to emission mitigation strategy happened from the initial phase of the time horizon. BOV 3-wheelers are chosen against the conventional 3-wheelers, and this option has made the choice very easy. This can be observed with increased share of BOV 3-Ws in Mumbai transport over the planning horizon.

References

Adebisi O (1985) Chapter 3: Transport in Northern Nigeria. In: Howe JDGF et al (eds) Rural transport in developing countries. Intermediate Technology Publications, London

Barwell IJ, Legget I (1986) Study on promotion of rural transport in Tanga Region, Consultancy Report by I.T. Transport Ltd. for GTZ on behalf of TIRDEP

Census of India (2001) Provisional population totals. Paper 1 of 2001

CMIE (2001) National income statistics-2001. CMIE, Mumbai

CPCB (1999) Parivesh, 6(1), Central Pollution Control Board, Ministry of Environment and Forests, Government of India, New Delhi

Gordon D (2005) Fiscal policies for sustainable best practices: international best practices. Report prepared for the Energy Foundation and The Hewlett Foundation

GTZ (2005) Training course on non-motorized transport, commissioned by Federal Ministry for Economic Cooperation and Development

Guitink P (1996) Strategic planning for non-motorized mobility, infrastructure notes. Transport No. OT-4, The World Bank

Guitink P, Holste S, Lebo J (1994) Non-motorized transport: confronting poverty through affordable mobility. Infrastructure notes. Transport No. UT-4, The World Bank

Heierli U (1993) Environmental limits to motorization. Niedermann, A.G, Switzerland

IPCC (1996) Climate Change 1995: The Science of Climate Change. Intergovernmental Panel on Climate Change; Houghton JT, Meira Filho LG, Callander BA, Harris N, Kattenberg A, Maskell K (eds). Cambridge University Press, Cambridge

Kaira CK (1985) Chapter 4: Transport in two Kenyan villages. In: Howe JDGF et al (eds) Rural transport in developing countries. Intermediate Technology Publications, London

Lowe M (1989) The bicycle: vehicle for a small planet. World Watch Institute Publications, Washington

MMRDA (Mumbai Metropolitan Regional Development Authority) (2011) Regional plan for Mumbai metropolitan development area. MMRDA, Mumbai

Pendakur SV (2000) A policy perspective for sustainable cities- non-motorized transport (NMT) in Asia, Urban Environmental Management – outreach series. CUC-AIT, Thailand

Rastogi R (2002) A policy sensitive behavioural model of transit access. PhD thesis, Department of Civil Engineering, Indian Institute of Technology, Bombay

Replogle M (1991), Non-motorized vehicles in Asian Cities. World Bank Technical report 162, - Washington, DC

The World Bank (1995) Non-motorized vehicles in ten Asian cities – trends, issues and policies, report TWU20. World Bank, Washington, DC

Yedla S, Shrestha RM, Anandrajah G (2005) Environmentally sustainable urban transportation – comparative analysis of local emission mitigation strategies vis-à-vis GHG mitigation strategies. Transp Policy, Elsevier 12(3):245–254

Chapter 9
Choosing Between Global Environmental Strategies Vis-à-Vis Local Mitigation Strategies: Application of Multi-criteria Analysis

9.1 Introduction

Transportation planning at large scale is predominantly influenced by the financial constraints and highly depends on external financial support. In the recent developments of global environmental issues, urban transportation research is more biased toward the GHG mitigation strategies. This resulted in increased inputs to reduce GHG emissions ignoring the local and more serious pollution. It is viewed in a way to get support from the developed world for infrastructure development by means of CDM and JI projects.

GHG mitigation strategy approach may ignore few potential alternative options like CNG technology, as they are not GHG friendly. This would lead to an unsustainable transportation development. Targeting global emission reduction alone cannot satisfy the local policy makers and users unless the local pollution is also reduced. At the same time transport sector is one which needs external support for development, and GHG mitigation strategies are found to be an attraction to involve the developed countries and development agencies. Therefore, there is a need for a logical base for the local policy makers in order for them to make a trade-off between these and go for the best bargains in urban transport sector.

With local demands on one side, lucrative financing offers coupled with international pressure from climate fraternity put the local administrators in a dilemma on which path to consider while they plan their transport system for the future years.

This chapter presents a case of analyzing the potential of both GHG mitigation strategies and local pollution (TSP and HC) mitigation strategies in reducing the nontarget emissions and also their economic efficiency. Such comparative analysis, which is missing in the literature, would support the decision on which mitigation strategies to follow to achieve environmentally sustainable transport system with least financial burden. It is a timely effort in those lines which can provide strong

Parts of this chapter are extracted from Yedla et al. (2005) and Yedla (2002).

S. Yedla, *Urban Transportation and the Environment*,
DOI 10.1007/978-81-322-2313-9_9

base for the policy makers to put forward the local interest on international platform without compromising on the interests of their counterparts.

9.2 Objectives of the Study

The objective of this case study is to carry out a comparative analysis between two different strategies for the development of sustainable urban transport system. It presents comparative analysis between GHG mitigation strategies and local pollutant emission strategies and their potential in reducing nontarget pollutants and potential saving in total transportation costs. This facilitates the policy makers with sufficient evidence to follow a particular strategy for a given set of constraints in any particular city (Yedla et al. 2005):

- Planning of Mumbai transport system to mitigate SPM and HC for a period of 20 years (1998–2020)
- Comparative analysis of transportation planning with respect to the GHG mitigation strategies and pollution control strategies to help policy makers identify correct strategies for transportation planning to achieve environmentally sustainable transportation systems.

Unlike Delhi, the transformation of transport sector toward energy-efficient and less polluting technologies has not picked momentum in Mumbai. This prompts for long-term action plan, which should come from the planning of transportation for a period of 20 years. Unlike Delhi where the pollution levels have reached threshold limits, Mumbai presents a case where the emissions would be high in the years to come. This resulted in increasing efforts to optimize the energy usage in transport sector and minimize the energy demands and also mitigation of emissions. However, this most expensive sector with a character of being highly nonpoint in nature needs huge funds to achieve any emission mitigation or energy efficiency. This prompts the need of involving the developed countries. For the developed countries to chip in into transport sector, GHG mitigation efforts are the prime attraction with possible CDM and JI projects in the infrastructure sector.

However, there is a trade-off between the interests of people who are more bothered about the local pollution and the policy makers who are more interested in getting support from developed countries to improve the ill-fated infrastructure. This trade-off needs research support to identify strategy which can bridge the gap with authentic confirmation that the interest of developed countries is met with that much of satisfaction to the public. The present study is aimed at making a comparison of transportation planning with respect to the GHG mitigation strategies and local pollution control strategies with emphasis on nontarget pollutant dynamics to help policy makers identify correct strategies for transportation planning to achieve environmentally sustainable transportation systems.

9.3 Methodology

9.3.1 Optimization Model

The dynamic optimization model presented in Chap. 8 is used for this analysis. The same is reproduced here to facilitate smoother reading.

9.3.1.1 Objective Function

To minimize total costs (capital, operational, and maintenance costs) of new vehicles and operating and maintenance costs of existing as well as new vehicles,

$$\sum_{i=1}^{I}\sum_{d=1}^{D}\sum_{v=1}^{V} X_{idv}(C_{idv} - S_{idv}) + \sum_{i=1}^{I}\sum_{d=1}^{D}\sum_{v=-v}^{t}\sum_{t=1}^{T} V_{idvt} * O_{idvt}$$

The total cost of transportation includes capital cost, operational and maintenance costs of the vehicles that should be added during the planning horizon, and operational and maintenance costs of the existing vehicles for the passenger transportation. All costs are expressed as a total net present value to the base year:

- *Variables*

  X_{idv} = number of vehicle, mode i, device d to be commissioned in year v
  V_{idvt} = km traveled by vehicle mode i, device d, vintage v, year t

- *Parameters*

  C_{idv} = Discounted capital cost of a vehicle mode i, device d, vintage v
  S_{idv} = Discounted salvage value of a vehicle mode i, device d, vintage v
  O_{idvt} = Operating cost of vehicle mode i, device d, vintage v, year t,

  The model includes the following constraints:

Travel demand constraint: The total travel service provided by existing and new vehicles in any year should be greater than or equal to the forecasted demand.

Vehicle capacity constraint: The total vehicle-km service provided by any type of vehicle should not exceed its maximum vehicle-km capacity of the total stock of that type of vehicle (i.e., existing and new units added).

Vehicle stock constraint: For candidate vehicles, total number of vehicles added to the transport system should not exceed the maximum limit on the number of vehicles that could be added during the planning horizon (which depends on maximum feasible penetration rate).

Emission constraint: Annual emission constraints – total emissions of the particular pollutant by all types of vehicles in a year should not exceed the target level of emission of that year. Overall emission constraints – total carbon dioxide emissions by all types of vehicles during the planning horizon should not exceed the target level and depend on overall emission reduction.

9.3.2 *Transportation Planning for Mumbai*

Transportation planning for Mumbai is done for the period of 1998–2020. Output of the model presents the vehicles added each year to cater the needs of transportation demand with a given set of constraints. Selection of different alternative transportation options depends on the emission constraints declared in the model. Emission mitigation targets influence the cost of total transportation system. This section presents the case of transportation planning for Mumbai without any emission mitigation targets.

9.3.2.1 Comparative Analysis of Emission Mitigation Strategies

The comparative analysis to reveal the strategic approach to the local emission mitigation against the GHG mitigation strategies is done in two steps. In the first step, the transportation planning for Mumbai transport system was done with restrictions on CO_2 emissions. Levels of CO_2 mitigation targets tried are 5, 10, 15, and 20 % of the overall CO_2 emissions over the period of 20 years. Under each level of CO_2 mitigation target, the vehicular mix and the emission of other pollutants are monitored. In the next step, the transportation planning was done with restrictions on total suspended particulate matter with different levels of emission mitigation at 5, 10, 15, and 20 % of the overall TSP being attempted. And the third step presents the similar analysis with emission restrictions on hydrocarbons (HC). Other parameters analyzed in all these three steps apart from the target pollutant include CO_2, SO_x, NO_x, TSP, and HC. Percentage change in each of these pollutants with varying levels of mitigation efforts is measured in all the three cases of emission mitigation.

9.4 Potential Candidates for Energy Efficiency and Pollution Mitigation

The following are the different alternative options that are available to improve the urban transport system. These options have a proven history of performance in different countries and cities.

In broad sense the alternative transportation options can be classified into two categories. One is alternative fuels and the other is alternative and advanced technologies and management practices. The following are the different individual candidates in each of those categories:

Alternative fuels

- Compressed natural gas (CNG) for cars/buses/three-wheelers/two-wheelers
- Electric and battery for three-wheelers/minibuses/cars
- Electricity for MRTS, trolley buses, etc.
- Duel fueled (hybrid)
- Fuel cells for cars
- Ethanol/methanol for cars
- Biofuel (biodiesel)
- LPG for cars/buses
- Fuel quality improvements (unleaded/low sulfur)

Technology and other options

- Four-stroke two-wheelers in place of two-stroke two-wheelers
- Control devices like magnetizers, catalytic converters, etc.
- Inspection and maintenance
- Increased share of public transport
- Efficient vehicles (as in developed countries)
- CVID (computer variable ignition device for cars)

In the present planning exercise, all modes of transport are selected, and the alternative options in the respective modes are chosen as candidate options for the optimal transport planning for Mumbai for the next 20 years. Selection of these alternatives is based on their energy-saving potential (ESP), emission reduction potential (ERP), and economic performance as presented in Chap. 4. The following are the alternative options chosen for the case of Mumbai:

Alternative option 1: Compressed natural gas (CNG) buses
Alternative option 2: Compressed natural gas (CNG) cars
Alternative option 3: Replacement of two-stroke two-wheelers by four-stroke two-wheelers
Alternative option 4: Compressed natural gas (CNG) three-wheelers
Alternative option 5: BOV three-wheelers

9.5 Transportation Planning for Sustainable Mobility

9.5.1 Transportation Planning for Mumbai Without Efforts for Emission Mitigation

The selected alternative options are used in the optimization model for the selection of best travel modes and devices to meet the travel demands at least cost. No

Table 9.1 Vehicular mix for Mumbai for the next 20 years (1998–2020)

Vehicles	Base year	2005	2010	2015	2020
Gasoline 3-wheeler	54,757	101,010	114,478	163,300	232,323
CNG 3-wheeler	0	0	0	0	0
BOV 3-wheeler	5,849	0	0	1	0
Gasoline car	145,461	254,895	340,069	655,613	833,383
Diesel car	36,365	166,243	189,195	18,619	14,895
CNG car	101,698	14,649	37,785	76,582	133,556
Diesel bus	6,966	14,647	19,432	24,184	33,461
CNG bus	0	0	0	0	0
2 stroke 2-wheeler	233,530	268,320	265,014	109,263	156,775
4 stroke 2-wheeler	20,866	225,508	468,245	1,087,894	1,534,209

emission mitigation is considered. Table 9.1 presents the vehicular mix for the next 20 years with a division of modes and devises.

The optimization model chose CNG cars over CNG three-wheelers and CNG buses. It is indeed an interesting observation to note that gasoline cars are continued to occupy a major share of vehicle stock. Battery-operated vehicles were selected in the beginning of the planning horizon. Diesel cars replaced by CNG cars and gasoline cars. However, the diesel buses are in rise leaving less emphasis on CNG buses. There is a clear jump in four-stroke two-wheelers over the two-stroke two-wheelers even without emission constraints. Total emissions of various pollutants and GHG under this vehicular mix are given in Table 9.2.

Over the planning horizon, NO_x emission increased by 5.5 times, whereas CO_2 and SO_x emissions increased by 4.4 and 4 times, respectively. TSP and HC increased by 3.3 times. This trend of emissions could be attributed to the increased share of CNG vehicles. Increased CNG usage would marginally increase the CO_2 emissions. However, this results in decreased TSP and HC emissions. Persistence of diesel buses contributed the higher growth of SO_x.

9.5.2 *Optimal Vehicular Mix for CO_2 Control Strategies*

The emphasis of many development projects on urban transportation research is on the greenhouse gas mitigation as a strategy to attract the participation of developed countries. The transport sector has been on a long debate for the possibility of it to qualify for CDM and JI projects. Developments in this direction gave substantial research inputs in assessing the potential of various technological and management options of transportation for the mitigation of GHG emissions. This section presents the planning of Mumbai transport system for various target levels of CO_2 emission mitigation and the corresponding changes in the vehicular mix and total emission levels of various local pollutants. Vehicular mix of Mumbai transport system under different levels of CO_2 mitigation targets are presented.

Table 9.2 Emissions of various pollutants over 20 years ('000 t)

Year	CO_2	TSP	SO_x	NO_x	HC
Base year	1,165.967	3.64	1.647	3.7645	28.09
2005	2,298.742	6.32	3.1404	8.909	51.508
2010	2,883.472	8.01	3.976	11.694	60.497
2015	3,498.261	7.92	5.012	15.896	76.331
2020	5,170.245	13.1	6.632	20.695	93.0362
Total	15,040.7	40.2	20.431	61.01	312.086

Increasing mitigation targets resulted in more penetration of alternative transportation options. Presence of BOV three-wheelers as an alternative option resulted in complete rejection of CNG options at all levels of CO_2 mitigation targets. Gasoline three-wheeler growth is nominal with multifold in BOV three-wheelers. The penetration is better in later part of the planning horizon. Gasoline cars improved in the early periods and remained unchanged in the later time period. Diesel cars did not grow much over the period of time. The stock of CNG cars did not change much with a certain level of technology penetration at the beginning of the planning horizon. Stock of diesel buses was found reducing over time with an increasing share of CNG buses. The induction of CNG technology in buses took place in the later part of the planning horizon.

Table 9.3 presents the total emission of various pollutants under different CO_2 mitigation targets. As it can be observed from the table, targeting at different levels of CO_2 mitigation resulted in considerable reduction in local pollution as well.

Figure 9.1 presents the trend of variations in other local pollutant emissions under the CO_2 mitigation targets. The change in emissions is shown as percentage change. SO_x showed a greater response to the CO_2 than any other emission. Higher level of CO_2 mitigation targets results in more efficient SO_x reduction. Trends of TSP and HC are similar as they share a relation as far as gasoline vehicles are concerned. NO_x showed little reduction. Though some variations are noticed in individual cases, the overall total emissions remained in a particular trend. And most of the emissions showed improvement. It is interesting to see that there is a considerable improvement in local pollution levels even with GHG mitigation strategies.

9.5.3 *Optimal Vehicular Mix for TSP Control Strategies*

This section examines the response of GHG emissions to the TSP mitigation targets. TSP has been selected as a candidate due to the fact that it is the most critical pollutant in urban transportation exceeding the allowable limits at most of the monitoring stations. Vehicular mix for the strategy of local emission control is examined using the optimization model.

Table 9.3 Total emissions of various pollutants over 20 years ('000 t) under different CO_2 mitigation targets

CO_2 mitigation target	CO_2	TSP	SO_x	NO_x	HC
Base	15,040.7	40.2	20.431	61.01	312.086
5 %	14,162.6	39.3379	21.042	60.501	309.736
10 %	13,550	37.4699	20.043	60.646	289.721
15 %	12,799.2	35.534	18.615	59.407	266.037
20 %	12,048.3	34.223	16.168	57.346	258.483

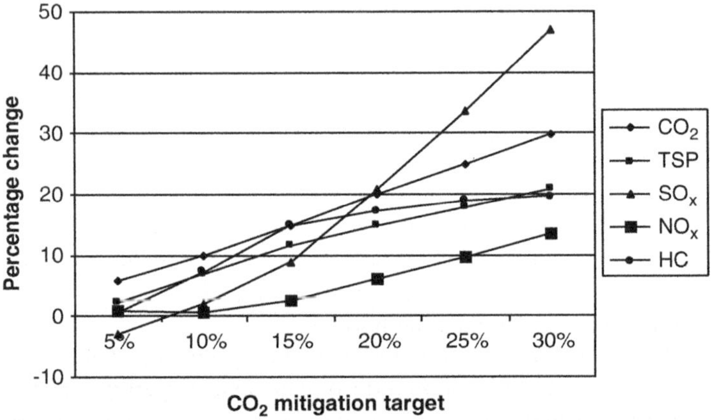

Fig. 9.1 Percentage change in pollutants under consideration at different levels of CO_2 mitigation strategies

Unlike the previous case of CO_2 mitigation strategies, TSP mitigation resulted in more shift toward the CNG technology and battery-operated vehicles. The vehicular mix also suggests that diesel cars and buses are least preferred. Unlike in the case of CO_2 emission mitigation strategies, CNG three-wheelers are selected in TSP strategies though it is at the higher level of mitigation targets. A similar trend is observed with CNG bus with increased share toward the later part of the time period.

The emission levels of various pollutants under different TSP mitigation targets are presented in Table 9.4. There is similar reduction in all emission levels over a range of TSP mitigation targets. It is interesting to note that TSP mitigation targets resulted in similar levels of mitigation in NO_x and SO_x. In fact SO_x showed better reduction levels than that of TSP. This could be due to the fact that more diesel vehicles are replaced by gasoline/CNG vehicles resulting in improved reduction of SO_x emission. HC showed less potential. CO_2 emissions showed considerable fall in the similar range of TSP mitigation targets. At 20 % CO_2 mitigation target, 15 % TSP reduction has been noticed, whereas at 20 % TSP mitigation target, 15.18 % CO_2 reduction could be achieved. However, in the latter case, the vehicular mix

Table 9.4 Total emissions of various pollutants over 20 years ('000 t) under different TSP mitigation targets

TSP mitigation target	TSP	CO_2	SO_x	NO_x	HC
Base	40.0	14,373.79	21.47	60.372	315.0299
5 %	38.0097	13,978.9	20.4808	60.879	306.8633
10 %	36.0616	13,398.64	19.4349	60.256	287.2807
15 %	40.0	14,373.79	21.47	60.372	315.0299
20 %	32.1653	12,191.79	16.5116	57.454	256.4233

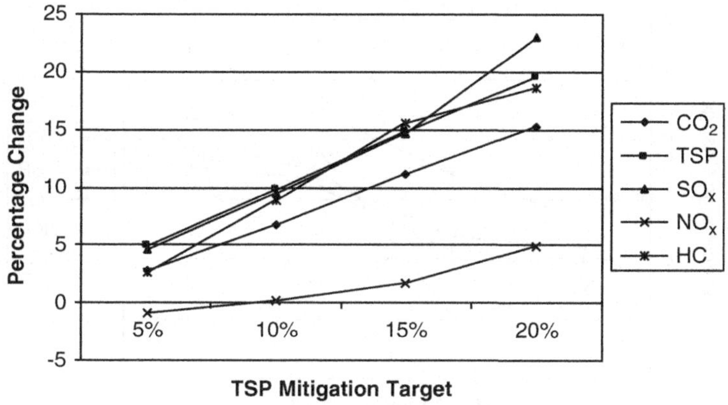

Fig. 9.2 Percentage change in pollutants under consideration at different levels of TSP mitigation strategies

differs from the former. SO_x performed slightly better in CO_2 mitigation strategies whereas others followed similar trends in both strategies (Fig. 9.2).

9.5.4 *Optimal Vehicular Mix for HC Control Strategies*

This section presents a case where hydrocarbon mitigation targets are set. This has been selected based on the fact that TSP mitigation targets mostly diesel vehicles and HC mitigation would target the replacement of gasoline vehicles. This would present a perfect comparison for efforts in different directions and the results of those efforts. Vehicular mix for Mumbai transport system with HC mitigation targets of 5, 10, 15, and 20 % is determined using the optimization model.

Four-stroke two-wheeler population increased considerably. CNG vehicles and battery-operated vehicles had a share in the vehicle stock but slightly on the latter half of the planning horizon. HC mitigation strategy presented a different case from the CO_2 and TSP mitigation targets. Table 9.5 presents the emission levels of various pollutants under different HC mitigation targets.

Table 9.5 Total emissions of various pollutants over 20 years ('000 t) under different HC mitigation targets

HC mitigation target	HC	CO_2	TSP	SO_x	NO_x
Base	302.2724	15,002.53	41.3	24.24048	58.6885
5 %	292.3995	15,039.22	41.28449	24.8764	58.1315
10 %	280.8776	15,017.27	41.0532	25.4029	57.60157
15 %	265.2733	13,907.07	38.9428	22.065	54.779
20 %	249.669	13,538.25	37.7566	21.38034	54.1781

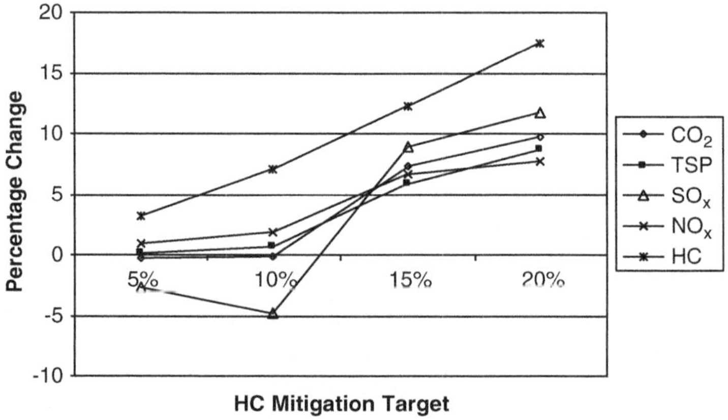

Fig. 9.3 Percentage change in pollutants under consideration at different levels of HC mitigation strategies

CO_2 emissions did not show much improvement with only 9.7 % reduction against the 15.3 % improvement in the case of TSP strategies. HC strategy was found to be doing well in controlling NO_x. It showed poor performance in reducing TSP emission with just 8.6 % reduction in TSP emission against the 15 % reduction in CO_2 mitigation targets. Figure 9.3 presents the percentage change in different pollutants with HC mitigation targets.

One distinct observation from this figure is that reduction of all nontarget emissions is much less than the target pollutant. At lower level of mitigation targets, the reduction of nontarget emissions is almost negligible except in the case of SO_x which showed an increasing trend.

Table 9.6 presents the total cost of transportation under all conditions of mitigation targets and different mitigation strategies. Cost of transportation in all three strategies is well within the limit and keeping very close to each other. Among the three strategies of emission mitigation, cost of transportation appears to be taking least value for the HC mitigation strategy closely followed by CO_2 and TSP mitigation strategies. HC mitigation strategy, which showed less potential in

Table 9.6 Total cost of transportation under different emission reduction targets and different mitigation strategies

	Total cost of transportation in billion USD		
Emission reduction target	CO_2 mitigation strategy	TSP mitigation strategy	HC mitigation strategy
0	4.8538	4.8693	4.8539
5	4.8623	4.8818	4.8598
10	4.8755	4.897	4.8658
15	4.8909	4.92	4.8758
20	4.9084	5.0266	4.9066

Table 9.7 Marginal abatement costs of CO_2 and TSP under CO_2 and TSP mitigation strategies

	MAC of CO_2 (USD/ton)			MAC of TSP (USD/ton)		
Mitigation target (%)	CO_2 strategy	TSP strategy	HC strategy	CO_2 strategy	TSP strategy	HC strategy
5	9.68	31.65	−160.81	9.53×10^3	6.42×10^3	10.4×10^4
10	14.55	28.41	−807.32	7.86×10^3	7.11×10^3	4.13×10^4
15	16.55	31.81	19.99	7.90×10^3	8.59×10^3	9.13×10^4
20	18.24	72.09	35.99	9.09×10^3	20.2×10^3	1.47×10^4

reducing nontarget emissions, showed better performance in economic terms. TSP mitigation strategy maintained the balance in both mitigation potential and economic performance.

9.5.5 *Marginal Abatement Cost*

Cost per ton of CO_2 mitigation revealed a different story altogether. Marginal abatement cost (MAC) of CO_2 under different strategies is presented in Table 9.7. With CO_2 mitigation strategy in urban transportation planning, cost per every ton of CO_2 reduction is in the range of 9.68–18 USD. MAC was found to be higher at higher mitigation targets. TSP reduction strategy showed higher MAC for CO_2 mitigation. However at moderate level of mitigation targets, MAC for CO_2 under TSP and CO_2 mitigation strategies is found to be close. The difference is increasing at higher mitigation targets. The difference between MAC values under different strategies is less between the mitigation target levels of 5–15 %.

Marginal abatement cost for the TSP mitigation is on higher side as the quantity of TSP mitigated is less compared to that of CO_2. As it can be observed from Tables 9.6 and 9.7, the difference between MAC of TSP is very less between CO_2 mitigation strategy and TSP mitigation strategy. This demonstrates the potential of local pollution mitigation strategies in handling the global emission mitigation. Therefore, employing local pollutant mitigation strategy in transportation planning would also cater the needs of GHG mitigation, which is a key factor in attracting international funding organization to invest in transport infrastructure development in developing countries. By employing the local pollutant emission mitigation strategies (LEMS) in urban transportation planning, it would be possible to handle both local and global pollutants from equal consensus with local policy makers and environmental activist and global actors.

References

Yedla S (2002) Environmentally sustainable urban transportation – comparative analysis of local mitigation strategies against GHG Mitigation Strategies, Post-doctoral fellowship report, AIT, Bangkok, Thailand

Yedla S, Shrestha RM, Anandrajah G (2005) Environmentally sustainable urban transportation – comparative analysis of local emission mitigation strategies vis-à-vis GHG mitigation strategies. Transp Policy, Elsevier 12(3):245–254